SAR

Side Effects and Drug Design

MEDICINAL RESEARCH
A Series of Monographs

Editor

Gary L. Grunewald, *University of Kansas*

Editorial Advisory Board
John W. Daly, *National Institutes of Health*
Gertrude B. Elion, *Wellcome Research Laboratories*
Saul B. Kadin, *Charles Pfizer and Co.*
Yvonne C. Martin, *Abbott Laboratories*
Wendel L. Nelson, *University of Washington*
P. S. Portoghese, *University of Minnesota*
David J. Triggle, *State University of New York at Buffalo*

Additional Volumes in Preparation

SAR

Side Effects and Drug Design

ERIC J. LIEN

University of Southern California
Los Angeles, California

CRC Press
Taylor & Francis Group
Boca Raton London New York

CRC Press is an imprint of the
Taylor & Francis Group, an **informa** business

CRC Press
Taylor & Francis Group
6000 Broken Sound Parkway NW, Suite 300
Boca Raton, FL 33487-2742

© 1987 by Taylor & Francis Group, LLC
CRC Press is an imprint of Taylor & Francis Group, an Informa business

No claim to original U.S. Government works

Visit the Taylor & Francis Web site at
http://www.taylorandfrancis.com

and the CRC Press Web site at
http://www.crcpress.com

To my parents, W. S. Lien and Y. C. Lin Lien, for their appreciation of the value of higher education

To my wife, Linda Lin-min, for her encouragement and understanding

To my sons, Raymond and Andrew, for letting me finish this seemingly never-ending project

To my students, for letting me use them as the test subjects of the materials presented

About the Series

This series of monographs was conceived by the late Dr. Fred Schueler of Tulane University. His untimely death came prior to the completion of the first volumes of the series, and his friend and very able colleague, Dr. Alfred Burger of the University of Virginia, took on the task of editing the first three volumes of the series. Dr. Burger's heavy responsibilities forced him to withdraw from further active participation with the series, and the next three volumes were selected from books already under contract to Marcel Dekker, Inc., with Dr. Gary Grunewald of the University of Kansas as Consulting Editor. At this point an Editorial Advisory Board was selected and Grunewald became the Series Editor. The traditions established under the leadership of Schueler and Burger will be continued.

It is hoped that books for the series will serve a useful role in the areas of medicinal chemistry, chemical pharmacology, and biochemistry. It is the intent of the Editor and the Editorial Advisory Board that books selected for the series should be timely and fill a definite need in the general areas mentioned. We welcome suggestions for future monographs in the series.

Gary L. Grunewald

Preface

This book is a combination of my research work in the area of the quantitative structure-activity relationship of disposition and activities of various pharmacological groups and a graduate course on SAR and drug design I have taught over the last 15 years. Also included is the first systematic attempt at linking side effects of drugs to their molecular structures and physicochemical properties.

This monograph is intended for graduate students in medicinal chemistry, pharmaceutical chemistry, pharmacology, and toxicology and for upper-class students in pharmacy, medicine, veterinary medicine, and other health sciences who are interested in learning about the strategy and problems associated with the design and development of new drugs, as well as for industrial research scientists in these fields.

No one has the gift of prophecy in designing new drugs, but it would be a costly proposition not to be familiar with existing working models, their advantages and weaknesses, and many reported cases of adverse reactions on clinically used drugs or common chemicals. It has been estimated that it will cost about $100 million to discover and market a new drug. It might cost even more money to discontinue a new drug and to settle all litigation, should the drug cause widespread serious problems, not to mention unnecessary human suffering.

I do not pretend to give many answers to these serious problems, but it is sincerely hoped that the reader will gain a broader perspective in selecting suitable models and appropriate parameters in correlation studies and know more about which potential side effects to look for during the tedious process of drug development.

I express sincere gratitude to my mentors, from whom I have learned to apply organic chemistry to medicine. The list includes the late Professor W. D. Kumler, the late Professor E. C. Jorgenson, Professor Corwin Hansch, and Professor E. J. Ariëns. My thanks also go to many collaborators and colleagues both within and outside the University of Southern California; without their support and assistance many of the studies included in this book would not have been possible. The hard work of many of my graduate students also made the preparation of this monograph less difficult.

The untiring professional support and assistance from the series editor, Professor Gary L. Grunewald; Ms. Sandra Beberman, Director, Editorial; Ms. Leslie Grundfest and Ms. Elaine Grohman, Production Editors; and their staff at Marcel Dekker, Inc., are deeply appreciated.

Eric J. Lien

Contents

SAR

Side Effects and Drug Design

One

Theories of Drug Action

Throughout the course of evolution and civilization, human beings have been seeking ways of altering and improving the states of the body and the mind. Numerous plants and natural products have been tried and have yielded many useful drugs. Subsequently, semisynthetic and synthetic compounds of varying complexity have been developed and marketed.

Molecular modification has been the major tactic for obtaining improved or new drugs since the turn of this century, and it is still a tremendously active field in modern drug research. It has been estimated the molecular modificaiton has produced in the last half century more potent and more useful drugs in various areas of therapeutics than have been reported in all of medical history (1). Many important concepts of drug-receptor interaction were also developed during this period of active reserach (see Table 1). As a consequence of this rapid proliferation, many "me-too" types of drugs and numerous varieties of therapeutic agents have appeared on the market. For example, there are over 1000 different drugs and many times as many preparations listed in the 1984 *Physician's Desk Reference* (2).

The large number of drugs imposes quite a challenge to physicians and pharmacists in choosing the best therapeutic agents available. For example, one may ask: Will two particular structurally related antibiotics or hormones be absorbed, distributed, and metabolized equally well? What types of sulfa drugs are most effective for a gastrointestinal or urinary tract infection? Which antihistamine should be used if one wants to avoid the central nervous system (CNS) depression side effect? Why should structurally unrelated drugs such as phenothiazines, haloperidol, and trimethobenzamide all produce the extrapyramidal syndrome?

Many of these types of questions can be rationalized in terms of the molecular structure and physicochemical properties of the drug molecules and the structural requirements of a given

biological system. The main purpose of this book is to give the
reader some current theories of drug action, examples of chemical manipulation of the molecular structure, and some quantitative and qualitative relationships between the molecular structure
and pharmacological activity. It is hoped that these examples
will contribute to a better understanding of drug action at the
molecular level. For research scientists involved in drug design,
the problem of side effects and toxicity cannot be ignored.
Often, an extremely promising compound may be abandoned, not
because of its lack of efficacy, but due to the manifestation of
a serious side effect in either experimental animals or human
subjects.

I. EVOLUTION OF CONCEPTS OF DRUG-RECEPTOR INTERACTION

Most of the important concepts and theories of drug action have
been formed in less than one century. It is reasonable to expect that more new discoveries will be made in the 1980s using
new technologies now becoming available in biology and chemistry
(see Table 1).

To understand the mechanism of drug action at the molecular
level, it is important to know the intermolecular forces that bind
drugs to their receptors. The determination of these forces by
experimental methods is very difficult. Nevertheless, on the
basis of facts already known, it is now generally accepted that
most structurally specific drugs attach themselves to their receptors through weak forces, including ionic forces, hydrogen bonding, and van der Waals forces. Only a few clinically useful
drugs are capable of forming strong covalent bonds with their

Table 1 Important Concepts in the Evolution of Drug-Receptor
Interactions and Drug Design

Year	Investigator(s)	New concept proposed or discovered
1898	P. Ehrlich	Haptophore (anchorer), toxophile (poisoner), receptor
1901	E. Overton, H. Meyer	Narcotic (depressant) activity and lipoid solubility
1904	C. Bohr, K. Hasselbach, et al.	Cooperative effect of hemoglobin
1913	L. Michaelis, M. L. Menten	Michaelis-Menten equation
1926	A. J. Clark, J. H. Gaddum	Occupancy theory in drug-receptor interaction
1935	U. S. von Euler	Prostaglandins
1937	R. Collander	Octanol/water partition coefficient and permeability
1939	J. Ferguson	Thermodynamic activity and narcosis
1940	L. P. Hammett	Hammett equation, linear free energy relationship
1951	A. Albert	Selective toxicity
1953	J. D. Watson, F. Crick	Double helical structure of DNA
1954	E. J. Ariëns	Intrinsic activity and affinity
1956	R. P. Stephenson	Efficacy
1957	B. B. Brodie, C. A. M. Hogben	pH-partition hypothesis
1957	A. Issacs, J. Lindenmann	Virus interference by interferons
1961	W. D. M. Paton	Rate theory
1961	M. Salame	Permacor method of predicting permeability through plastics

Table 1 (Continued).

Year	Investigator(s)	New concept proposed or discovered
1962	C. Hansch, T. Fujita, et al.	π constant, the Hansch multiple regression analysis approach in structure-activity relationship
1964	D. E. Koshland, Jr.	Induced-fit theory
1964	B. Belleau	Macromolecular perturbation theory
1965	J. Monod, P. Changeux, et al.	Allosteric site
1967	B. R. Baker	Active-site-directed irreversible enzyme inhibitors
1970	H. M. Temin, S. Mitzutani, D. Baltimore	RNA-dependent DNA polymerase (reverse transcriptase)
1972	G. A. Robinson, E. W. Sutherland, et al.	Cyclic-AMP as the second messenger
1977	G. Weissman	Cyclic-AMP and cyclic-GMP as the (+) and (−) mediators
1980	C. Milstein	Monoclonal antibodies
1980	P. Berg	Construction of the first recombinant DNA
1984	L. Montagnier	Lymphadenopathy-associated virus (LAV)[a]
1984	R. Gallo	Human T-cell leukemia/lymphoma virus (HTLV)
1984	R. Gallo	AIDS virus (HTLV-III)[a]

[a]The International Committee on the Taxonomy of Viruses has recommended the name Human Immunodeficiency Virus (HIV) for these viruses. [Coffin, J., et al. (1986). *Science 232:* 697.]

receptors. The best examples are anticancer alkylating agents and acetylcholinesterase inhibitors such as neostigmine, physostigmine, and organophosphates [e.g., diisopropyl fluorophosphate (DFP)]. The latter action is due to the irreversible inhibition of acetylcholinesterase through covalent bond formation (3). This involves the formation of the corresponding carbomoyl acetylcholinesterase and the phosphorylated enzyme, respectively.

Many irreversible inhibitors of enzymes have been developed by Baker and co-workers employing the concept of active-site-directed irreversible inhibitor operating by a covalent linkage with an enzymatic nucleophillic group situated outside the active site (4). Some compounds with interesting activities have emerged; however, no clinically useful drugs have yet resulted from this approach.

Because different types of drugs possessing different functional groups are capable of having different types of interactions with various receptors, several theories of drug action have been proposed and modified by several investigators. The essence of these theories will be discussed here. For a more comprehensive treatment of this subject, the reader should refer to Korolkova's monograph (5).

II. OCCUPANCY THEORY

Clark (6) and Gaddum (7) proposed that the intensity of pharmacological effect is directly proportional to the number of receptors occupied by the drug molecules. According to this theory, drug-receptor interactions closely follow Langmuir's adsorption isotherm (8). The number of occupied receptors depend on the concentration of the drug in the receptor compartment

(the biophase) and on the total number of receptors. Maximal
action will correspond to the occupation of all receptors. The
theory of occupancy is valid only when the drug-receptor interaction is identical for all drugs under consideration (i.e., all
of them can produce the same maximal response). However,
certain drugs, such as acetylcholine congeners, never elicit
maximal response no matter how high the concentration is.
Ariëns (9) and Stephenson (10) proposed modifications of this
theory to take into account this and other discrepancies. They
proposed that two steps are involved in drug action: (a) combination of drug and receptor, and (b) production of effect.
Thus any drug may have structural features that contribute in-
dependently to the *affinity* for the receptor and to the *intrinsic
activity* (*efficacy*) with which the drug-receptor combination
initiates the response. The Ariëns-Stephenson concept retains
the assumption that the response is related to the number of
drug-receptor complexes. Compounds of low intrinsic activity
and high affinity may thus be used as antagonists; compounds
of high affinity and high intrinsic activity work as agonists.

III. RATE THEORY

Paton and co-workers (11–13) proposed that the activation of
receptors is proportional not to the number of occupied receptors
but to the rate of encounters of the drug with its receptor. In
other words, a stable drug-receptor complex does not necessarily
yield a maximal effect, and the pharmacological activity is a
function only of the rate of association and dissociation between
the drug molecules and the receptor. According to this theory,
an agonist would have a fast rate of association and dissociation

and would produce several impulses per unit of time, much like
a ping-pong ball bouncing back and forth. On the other hand,
an antagonist would have a fast rate of association but a slow
rate of dissociation.

Both the occupancy and the rate theories do not have a
plausible physicochemical basis for the interpretation of phe-
nomena involving receptors at the molecular level. Furthermore,
the rate-limiting step may not be the drug-receptor binding
step. For example, it has been proposed (14) that the action
of atropine is limited by the rate of access to the receptor
region, not by the rate of the atropine-receptor interaction.

IV. INDUCED-FIT THEORY: ALLOSTERIC SITE AND COOPERATIVE EFFECT

The classical "key-lock" or "template" hypothesis was not ade-
quate to explain many observed facts of substrate-enzyme interactions. Koshland (15) suggested that the active site of an
enzyme does not necessarily need to have a morphology comple-
mentary to that of the substrate. The concept of a flexible
enzyme has been found to be more useful than the concept of a
rigid protein. The induced-fit hypothesis advanced by Koshland
(16, 17) suggests that the substrate induces a change in the
enzyme conformation after binding with the enzyme takes place;
this may orient the catalytical groups in a favorable way for the
subsequent reaction. Increased or decreased substrate size
may prevent proper alignment of the catalytical groups, and
thus leads to inactivation of the enzyme. Even though the induced-fit concept was proposed to explain enzymic action on substrate, a similar enzyme-activating or enzyme-deactivating effect

can be produced by a drug. This effect may be produced by
complexation of a drug with the enzyme's normal substrate and
subsequently increases or decreases its volume. It may also
modify the enzyme action by binding at an allosteric site. Monod,
Changeux, and Jacob (18) originally coined the term "allosteric"
to designate a topologically separate site that lacks catalytic
activity but can exert regulatory control over the active site.
Some enzyme inhibitors, especially the ones that bear no
structural relationship with the normal substrate, may exert
their effect at allosteric sites.

The varying roles of a hormone activator (see Section VI)
may be explained on the basis of such a molecular mechanism.
That is, the hormone may act to overcome the competitive effect
of an inhibitor (the hormone acts as an activator only in the
presence of the inhibitor). Alternatively, the hormone or activator may compete directly with the competitive inhibitor for
the inhibiting site. In this case, the hormone should be structurally similar to the inhibitor, even though it has no similarity
to the substrate itself.

The *cooperative effect* (19) was initially discovered in
hemoglobin and appears to be a characteristic of many enzymes
composed of subunits. When a plot of fractional saturation of
oxygen binding sites versus oxygen concentration is made, a
sigmoid curve is obtained in contrast to the usual hyperbolic
curve of Michaelis-Menten kinetics. Such a curve indicates
that the first molecule of ligand (O_2) somehow makes it easier
for the second molecule to bind. This phenomenon was termed
"cooperative" and is today more accurately called a "positive
homotropic effect." Both electrostatic interaction and conforma-

tional explanation have been proposed to explain such a phenomenon (20).

V. MACROMOLECULAR PERTURBATION THEORY

In 1964, Belleau proposed the macromolecular perturbation theory of drug action (21). From the thermodynamic parameters for the binding of homologous series of quaternary ammonium compounds by acetylcholinesterase (22, 23), Belleau reasoned that in the drug-receptor interaction two types of perturbation can occur on the macromolecule-drug complex: (a) specific conformational perturbation or specific ordering caused by an agonist, which makes possible the adsorption of certain molecules related to the substrate, and (b) nonspecific conformational pertubation or disordering caused by an antagonist; this type of nonspecific perturbation may serve to accommodate extraneous molecules. In either case the hydrophobic interaction and the structure of the water molecules surrounding the receptor site are quite important. In the case of the partial agonist or antagonist, the drug has both characteristics; that is, it contributes to specific as well as nonspecific perturbation, resulting in a mixture of two complexes. Further experimental data have been obtained to lend more support to this theory (24).

In view of the fact that stricter structure-activity relationships exist for histaminelike drugs than for antihistaminics (25), one can visualize that antihistaminics probably occupy not only the receptor area of histamine but also the accessory hydrophobic area of the histamine receptor (26, 27); alternatively, they may inhibit the histamine action by molecular perturbation mechanism, just like anticholinergics on a cholinergic receptor (23).

VI.   THE SECOND MESSENGER CONCEPT

In recent years it has become clear that the receptors with which
many hormones combine are closely related to or may even be
part of an adenyl cyclase system (28). For example, catecholamines, glucagon, adenocorticotropic hormone (ACTH), luteiniz-
ing hormone (LH), vasopressin, parathyroid hormone, thyrotro-
pin-releasing hormone (TRH), thyroid-stimulating hormone (TSH),
melanocyte-stimulating hormone (MSH), serotonin, histamine,
and so on, have been shown to affect the level of cyclic adeno-
sine monophosphate (AMP). Depending on whether the cyclase
is stimulated or inhibited, the result of the hormone-receptor
interaction will be an increase or decrease in the intracellular
level of cyclic AMP. When the stimulus is an increase in the
cyclic AMP level, the hormone may be regarded as a first
messenger, with cyclic AMP as the second messenger. This is
represented schematically in Figure 1.

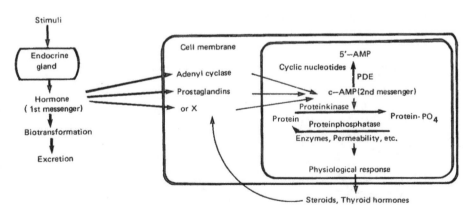

Figure 1 Relationship between the action of a hormone (the first
messenger) and that of cyclic-AMP (the second messenger).

It is known that the methylxanthines (e.g., caffeine, theophylline) are competitive inhibitors of cyclic nucleotide phosphodiesterase, an enzyme that catalyzes the conversion of cyclic AMP to 5'-AMP. Cyclic AMP concentrations are thus elevated in some tissues following exposure to these drugs.

In view of the fact that prostaglandins (PGs) also affect cyclic AMP in many biological systems and that PGs are synthesized in cell membrane (29), it is quite possible that prostaglandins also use cyclic AMP as the second messenger in various organs and tissues. It has been suggested that PGs serve as a negative-feedback mechanism in the regulatory process (29). Prostaglandins have been shown to induce fever and inflammation, and the recent findings of the inhibition of prostaglandin release by aspirinlike drugs (30-32) account for their anti-inflammatory and antipyretic activity. Discovery of better prostaglandin synthetase inhibitors may possibly lead to more effective agents for the treatment of arthritis (33). Other applications of prostaglandins, such as fertility control and improvement in blood banking, appear to be promising (33).

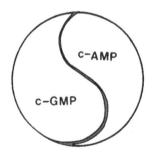

Mediator Release

Secretion decreased by cyclic-AMP
      increased by cyclic-GMP

E.g.: Lysosomal enzymes

       Histamine

       SRS-A

       Lymphokines

Figure 2   The balancing effects of cyclic-AMP and cyclic-GMP resemble the Ying-Yang hypothesis of Chinease medicine.

It has further been suggested that cyclic-AMP and cyclic guanylic acid (GMP) may have opposing effects to balance the release of mediators in inflammation (34) fitting the ancient Ying-Yang hypothesis of Chinese medicine (Figure 2).

VII. ACTIVE-SITE-DIRECTED IRREVERSIBLE INHIBITORS

B. R. Baker has designed many active-site-directed irreversible enzyme inhibitors; both the endo and the exo mechanisms have been explored (35). Among the several hundred compounds studied, the following triazine inhibitor of dihydro folate reductase has shown some promise in clinical trials in cancer chemotherapy (1):

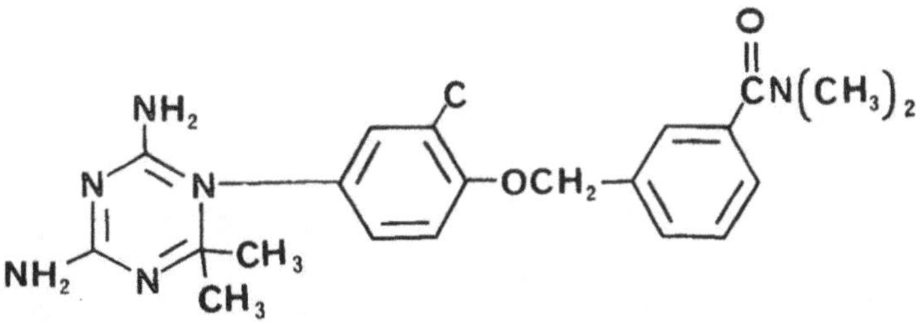

(1)

Hansch and co-workers have analyzed the structure-activity relationship of this series of compounds and that of quinazolines (36). A map of the area around the binding site has been proposed as follows (2):

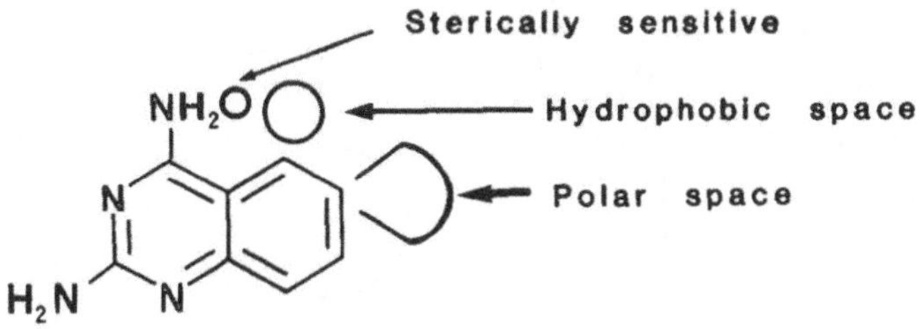

(2)

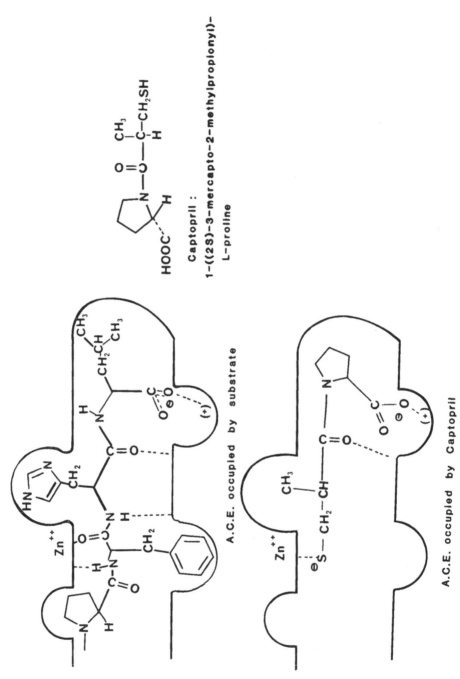

Captopril :
1-((2S)-3-mercapto-2-methylproplonyl)-
L-proline

A.C.E. occupied by substrate

A.C.E. occupied by Captopril

Scheme 1

VIII. MECHANISM- AND PHARMACOPHORE-BASED DRUG DESIGN

Captopril, a synthetic derivative of proline, is an oral antihypertensive based on the inhibition of angiotensin-converting enzyme (ACE) (Scheme 1, page 14). This new drug is able to bind to the enzyme as a specific competitive inhibitor, thus preventing the conversion of angiotensin I to angiotensin II (37, 38) and leading to its antihypertensive effect (Scheme 2).

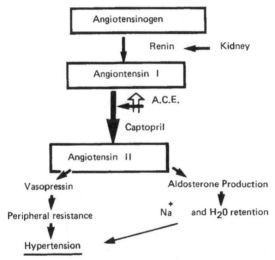

Scheme 2 Mechanism of action of captopril as an antihypertensive agent.

Verapamil HCl has been marketed in the United States in 1981 (39) as an antiarrhythmic agent. It owes this activity to the inhibition of excitation-contraction coupling in mammalian myocardium and vascular smooth muscle (40, 41). Many β-adrenergic

---

Scheme 1 Schematic representation of the ACE occupied by captopril as compared to the same enzyme occupied by the natural substrate.

Table 2 Structural Formulas of β-Adrenergic Drugs[a]

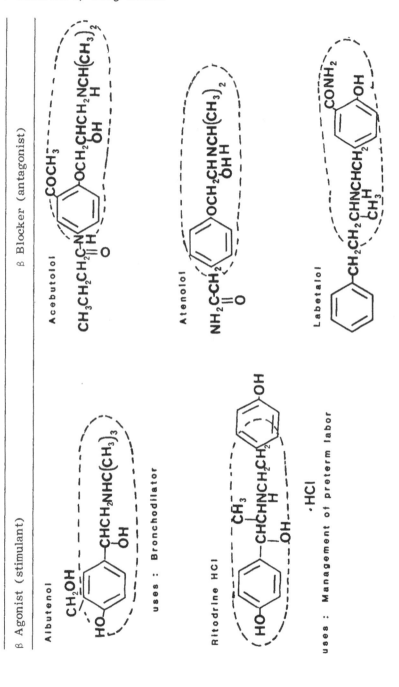

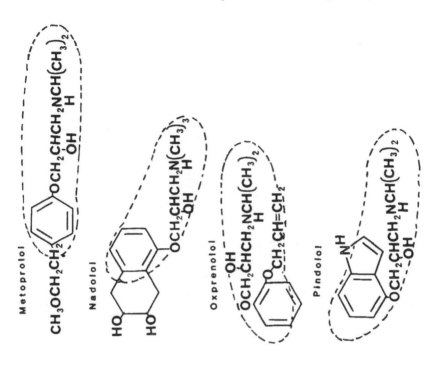

Metoprolol

Nadolol

Oxprenolol

Pindolol

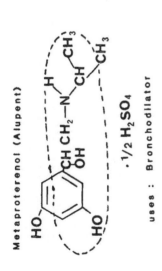

Metaproterenol (Alupent)

· ½ H₂SO₄

uses : Bronchodilator

Table 2 (Continued).

β Agonist (stimulant)	β Blocker (antagonist)

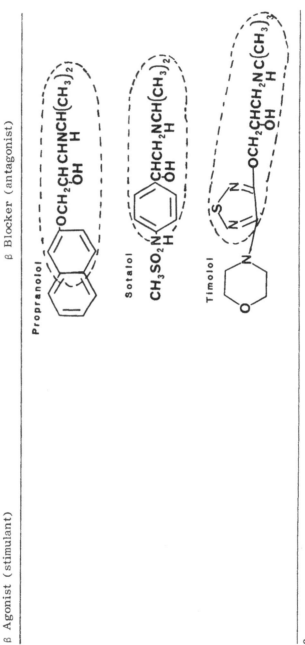

Propranolol

$OCH_2CH\ CHNCH(CH_3)_2$
$\quad\quad\quad\overset{|}{OH}\ \overset{|}{H}$

Sotalol

$CH_3SO_2N\quad\quad CHCH_2NCH(CH_3)_2$
$\quad\quad\overset{|}{H}\quad\quad\overset{|}{OH}\ \overset{|}{H}$

Timolol

$OCH_2CHCH_2NC(CH_3)_3$
$\quad\quad\overset{|}{OH}\quad\overset{|}{H}$

aThe pharmacophore for the β receptor is encircled for easy identification.

Table 3 Physicochemical Properties and Some Pharmacokinetic Parameters of β-Adrenergic Blockers[a]

Drug name	log P'	log K_p	log Cl_{NR} (ml/min)	log Cl_R (ml/min)	log V_{uss} (L)
Penbutolol	1.91	2.97	—	—	—
Bufuralol	1.80	2.28	2.73	0.60	3.16
Tolamolol	1.52	2.28	2.86	1.57	3.21
Propranolol	1.13	2.40	2.83	1.43	3.29
Alprenolol	0.98	1.78	2.64	1.26	2.50
Oxprenolol	0.20	2.34	2.24	1.34	—
Acebutolol	-0.21	0.82	2.68	2.47	2.10
Timolol	-0.23	1.45	2.72	1.85	1.97
Metoprolol	-0.30	1.12	2.85	1.97	2.38
Pindolol	-0.39	1.44	2.42	2.43	2.30
Practolol	-1.30	0.16	—	—	—
Atenolol	-1.70	-0.23	1.00	2.23	1.90
Nadolol	-1.70	0.87	1.69	2.18	2.27
Sotalol	-2.00	—	—	2.20	1.98

[a]P', Octanol/buffer (pH 7.4) by shake flask method; K_p, plasma protein/plasma water partition coefficient; Cl_{NR}, nonrenal clearance; Cl_R, renal clearance; V_{uss}, steady-state volume of distribution referenced to the unbound drug in plasma.
Source: Adapted from Ref. 43.

blockers have been marketed for the treatment of angina, hypertension, patients after heart attack, and glaucoma. They all have a secondary amino group attached to 2-propanol—therefore, the generic names with an ending of -olol (e.g., acebutolol, labetalol, metoprolol, propanolol, timolol, etc.) (42) (see Table 2). β-Adrenergic agonists such as albutenol, on the other hand, has been promoted as a bronchodilator, while ritodrine HCl is for the management of preterm labor (see Table 1). These applications are due primarily to the greater affinity and intrinsic activity of the drug to the β_2-adrenergic receptors at the respective target organs.

Table 3 summarizes the important physicochemical properties and the pharmacokinetic parameters of some β blockers. The structure-pharmacokinetics relationship of these drugs have been reported by Hinderling et al. (43).

REFERENCES

1. Schuler, F. W. (1964). Perspective in drug therapy, in R. F. Gould, Ed., *Molecular Modification in Drug Design*, Advances in Chemistry Series 45. American Chemical Society, Washington, D.C., pp. 221–222.

2. Huff, B. B., Ed. (1984). *Physicians' Desk Reference*, 38th ed. Medical Economics, Oradell, N.J.

3. Gisvols, O. (1968). Cholinergic agents and related drugs, in C. O. Wilson, O. Gisvold and R. F. Doerge, Eds., *Textbook of Organic Medicinal and Pharmaceutical Chemistry*, 5th ed. J. B. Lippincott, Philadelphia, pp. 453–467.

4. Baker, B. R. (1967). *Design of Active-Site Directed Irreversible Enzyme Inhibitors*. Wiley, New York,

5. Korolkovas, A. (1970). *Essentials of Molecular Pharmacology*. Wiley-Interscience, New York.

6. Clark, A. J. (1926). The reaction between acetylcholine and muscle cells. *J. Physiol. 61:* 530. The antagonism of acetylcholine by atropine. *Ibid., 61:* 547.

7. Gaddum, J. H. (1926). The action of adrenaline and ergotamine on the uterus of rabbit. *J. Physiol. 61:* 141. (1937). Quantitative effects of antagonistic drugs-discussion. *Ibid., 89:* 7.

8. Langmuir, I. (1918). The adsorption of gases on plane surfaces of glass, mica and platinum. *J. Am. Chem. Soc. 40:* 1361.

9. Ariëns, E. J. (1954). Affinity and intrinsic activity in the theory of competitive inhibition: I. Problems and theory. *Arch. Int. Pharmacodyn. Ther., 99:* 32.

10. Stephenson, R. P. (1956). A modification of receptor theory. *Brit. J. Pharmacol. 11:* 379.

11. Paton, W. D. M. (1961). A theory of drug action based on the rate of drug-receptor combination, *Proc. R. Soc. London, Ser. B., 153:* 21.

12. Paton, W. D. M., and Rang, H. P. (1966). A kinetic approach to the mechanism of drug action. *Adv. Drug Res. 3:* 57.

13. Paton, W. D. M., and Payne, J. P. (1968). *Pharmacological Principles and Practice.* J. & A. Churchill, London.

14. Thron, C. D., and Waud, D. R. (1968). The rate of action of atropine. *J. Pharmacol. Exp. Ther. 160:* 91.

15. Koshland, D. E., Jr. (1958). Application of a theory of enzyme specificity to protein synthesis. *Proc. Natl. Acad. Sci., U.S.A. 44:* 98.

16. Koshland, D. E., Jr. (1964). Conformation changes at the active site during enzyme action. *Fed. Proc. 23:* 719.

17. Koshland, D. E., Jr. (1968). The molecular pharmacology of anesthesia—discussion. *Fed. Proc. 27:* 907.

18. Monod, J., Changeux, P., and Jacob, F. (1965). Allosteric proteins and cellular control systems. *J. Mol. Biol. 6:* 306.

19. Bohr, C., Hasselback, K., and Krogh, A. (1904). Ueber einen in biologischer Beziehug wichtigen Einfluss, den die Kohlensäurespannung des Blutes auf dessen Sauerstoffbindung übt. *Skand. Arch. Physiol. 16:* 402.

20. Koshland, D. E., Jr., and Neet, K. E. (1968). The catalytic and regulatory properties of enzymes. *Annu. Rev. Biochem.* 37: 359.

21. Belleau, B. (1964). A molecular theory of drug action based on induced conformational perturbation of receptors. *J. Med. Chem.* 7: 776.

22. Belleau, B. (1965). Conformational perturbation in relation to the regulation of enzyme and receptor behavior. *Adv. Drug Res.* 2: 89.

23. Belleau, B. (1967). Water as the determinant of the thermodynamic transitions in the interaction of aliphatic chains with the acetylcholinesterase and the cholinergic receptors. *Ann. N.Y. Acad. Sci.* 144: 705.

24. Belleau, B., and Lavoie, J. L. (1968). A biophysical basis of ligand-induced activation of excitable membranes and associated enzymes. A thermodynamic study using acetyl-cholinesterase as a model receptor. *Can. J. Biochem.* 46: 1397.

25. Ariëns, E. J., and Simmonis, A. M. (1966). Aspetti Della Farmacologic Molecolare. *Farmaco (Pavia)* 21: 581.

26. van den Brink, F. G., and Lien, E. J. (1977). pD_2, pA_2 and PD_2'-values of a series of compounds in a histaminic and a cholinergic system. *Eur. J. Pharmacol.* 44: 251.

27. van den Brink, F. G., and Lien, E. J. (1977). Competitive and noncompetitive antagonism, in M. R. e Silva, Ed., *Handbuch der experimentellen pharmakologie*, Vol. XVIII/2, Section B. Springer-Verlag, Berlin, pp. 333−367.

28. Robinson, G. A., Butcher, R. W., and Sutherland, E. W. (1972). *Cyclic AMP*. Academic Press, New York, p. 22.

29. Pike, J. E. (1971). Prostaglandins. *Sci. Am.* 225: 84.

30. Vane, J. R. (1971). Inhibition of prostaglandin synthesis as a mechanism of action for aspirin-like drugs. *Nature New Biol.* 231: 232.

31. Smith, J. B., and Willia, A. L. (1971). Aspirin selectively inhibits prostaglandin production in human platelets. *Nature New Biol.* 231: 235.

32. Bethel, R. A., and Eakins, K. E. (1971). Antagonism by polyphloretin phosphate of the intraocular pressure rise induced by prostaglandin and formaldehyde in the rabbit eye. *Fed. Proc.* 30: 626.

33. Lien, E. J. (1974). Potential applications of prostaglandins and antiprostaglandins. *Calif. Pharm. 22*(3): 22–35; *22*(4): 24–34.

34. Vane, J. R., Weissman, G., and Zurier, R. B. (1977). *Pain and Prostaglandins—New Clinical Perspectives.* Science and Medicine Publishing Co., Inc., a Wellcome Medical Education Service, Research Triangle Park, NC.

35. Baker, B. R. (1967). *Design of Active-Site-Sirected Irreversible Enzyme Inhibitors.* Wiley, New York.

36. Fukunaga, J. Y., Hansch, C., and Steller, E. E. (1976). Inhibition of dihydrofolate reductase. Structure-activity correlations of quinazolines. *J. Med. Chem. 19*: 605.

37. Cushman, D. W., and Ondetti, M. A. (1980). Inhibitors of angiotensin-converting enzyme for treatment of hypertension. *Biochem. Pharmacol. 29*: 1871.

38. Riley, T. N., and Fischer, R. G. (1981). Report on new drugs—1981. *U.S. Pharm. 1931,* August, 78–88.

39. Hussar, D. A. (1982). New drugs of the year. *Am. Pharm. NS22*(3): 29–47.

40. Fleckenstein, A. (1977). Specific pharmacology of calcium in myocardium, cardiac pacemakers, and vascular smooth muscle. *Annu. Rev. Pharmacol. Toxicol. 17*: 49.

41. Flaim, S. F., and Zelis, R. Ed. (1982). *Calcium Blockers.* Urban & Schwarzenberg, Baltimore, Md.

42. Sanders, H. J. (1982). New drugs for combating heart disease, *Chem. Eng. News.,* July, 24–38.

43. Hinderling, P. H., Schmidlin, O., and Seydel, J. K. (1984). Quantitative relationships between structure and pharmacokinetics of beta-adrenoceptor blocking agents in man. *J. Pharmacokinet. Biopharm. 12*: 263–287.

Two

Improvement of Therapeutic Agents by Molecular Modification

I. MODIFICATION OF THE ABSORPTION OF DRUGS

If a drug is intended for systemic action, it has to be absorbed and properly distributed from the site of administration to the site of action. It is well established that most organic drug molecules are absorbed from the gastrointestinal (GI) tract by by passive diffusion. Therefore, they follow the generalizations of the pH-partition theory (1–10). Namely, in the GI tract the drug must not dissociate completely, and it must have a large enough lipoid/water partition coefficient to cross the lipoprotein membrane. Penetration of drugs into cerebrospinal fluid is also known to be highly dependent on the lipoid/water partition coefficient (11).

In a homologous series of phenoxymethylpenicillin (penicillin V) derivatives, it has been found that the serum concentration after oral administration of 0.3 g of the drug increases as the side chain increases from H to C_5 (12) (Table 4).

Table 4 Structure-Serum Concentration Relationship of Penicillin V Derivatives[a]

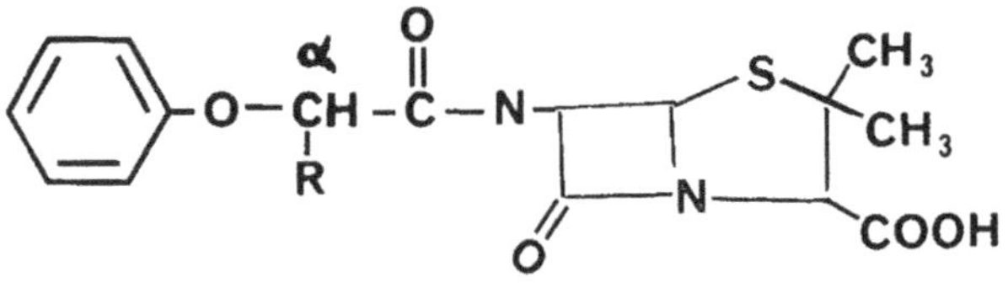

$\mu g/ml$		
	R	Penicillin
2.09	H	Phenoxymethyl
3.02	C_2H_5	α-Phenoxy-*n*-propyl-
6.69	$n\text{-}C_3H_7$	α-Phenoxy-*n*-butyl-
7.07	$i\text{-}C_3H_7$	α-Phenoxy-*i*-butyl-
8.32	$n\text{-}C_4H_9$	α-Phenoxy-*n*-amyl-
10.13	$n\text{-}C_5H_{11}$	α-Phenoxy-*n*-hexyl-

[a]Human serum concentration 1 hr after oral administration.
Source: Ref. 12.

The increase in GI absorption is apparently due to the increased lipoid/water partition coefficient, since each CH_2 will increase the log $P_{octanol/water}$ value by 0.5 and an *n*-amyl group will increase the log P value by 2.5 (or increase the P 315-fold). Another example of achieving a higher serum level of the parent drug by making easily hydrolyzed derivatives involves a series of acycloxymethyl esters of ampicillin. Higher serum levels of ampicillin in normal volunteers following oral administration of pivaloyloxymethyl ester of ampicillin has been found by Daehne and co-workers (13) (3).

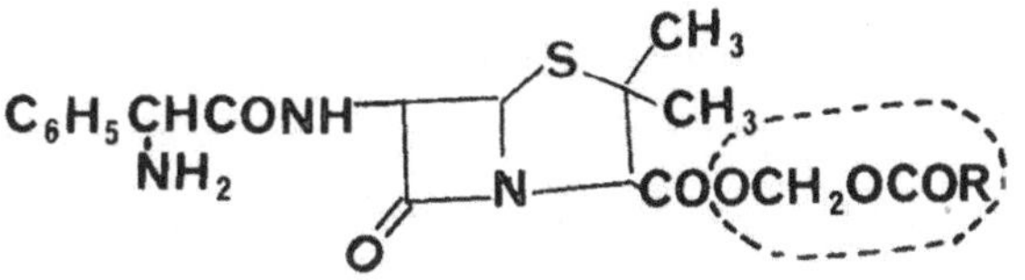

$$C_6H_5CHCONH \underset{NH_2}{\overset{}{|}} \qquad (3)$$

Acycloxymethyl esters of D-α-aminobenzylpenicillin (ampicillin)

Pivaloyloxymethyl ester; R = $C(CH_3)_3$

The pivaloyloxymethyl ester is absorbed more efficiently from the GI tract than ampicillin itself, and it is rapidly hydrolyzed to ampicillin, with the result that high blood and tissue levels of the parent drug are attained (13).

II. OPTIMIZATION OF TIME COURSE AND DISTRIBUTION BY CHEMICAL MODIFICATION

There are numerous examples of chemical modifications employed to extend the duration of drug action, such as protamine zinc insulin, testosterone propionate, procaine G penicillin, hydroxy-

progesterone caproate, triamcinolone hexacetonide, and hydrocortisone cypionate (21-cyclopentanepropionate ester of hydroxy-
cortisone). Most of these preparations achieve their longer
duration of action by virtue of their decreased solubility in
water and slower rate of dissolution.

On the other hand, by making a derivative of a smaller
acid salt or polyfunctional salt, one can increase its water
solubility. A few examples of this category are the following:
chloramphenicol sodium succinate has a greater solubility in
water than the parent compound and it is more rapidly absorbed by intravenous injection; a mixture of β-methasone
sodium phosphate and sodium acetate and hydrocortisone sodium
succinate are used for rapid action of these steroids by injection.

Once in the bloodstream, the accumulation of the drug in
various tissues is determined primarily by its physicochemical
properties, such as the lipoid/water solubility and the acid-base
strength. The best known examples of selective distribution are
the various Roentgen-contrast agents used for x-ray examination
of the ureter pelvis of the kidney (pyelography) or for examination of the gallbladder (cholecystography). The carrier moieties
for these purposes are compounds that are excreted rapidly by
the kidney into the urine or by the gallbladder into the bile
duct. *Para*-aminohippuric acid is an excretion product from *p*-
aminobenzoic acid. Because it is an organic acid of high water
solubility, *p*-aminohippuric acid is excreted in the urine by
ultrafiltration and by an active transport system.

Active secretion gives practically complete clearance of the
blood passing the nephrons for that compound. Owing to the
high degree of dissociation of the acid, there is no renal reab-
sorption. *Para*-aminohippuric acid has been used as a carrier for

iodine, a heavy element with a high density for the absorption of x-rays. Following this compound as a lead, drug designers have developed a variety of organic acids with a high renal clearance as carriers for iodine in radiopaques of pyelography. A few drugs used clinically for this purpose are (4–7)

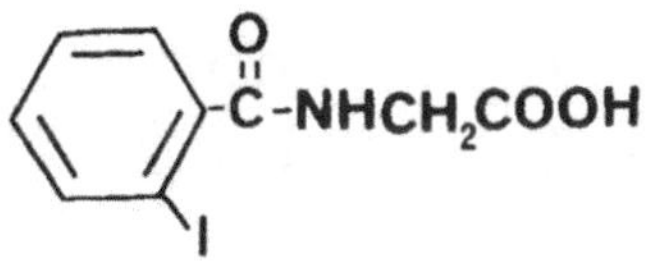

Hippuran

(4)

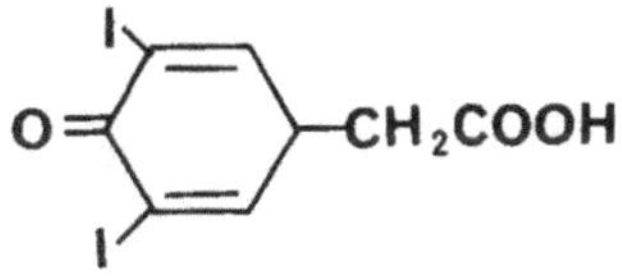

Diodone

(5)

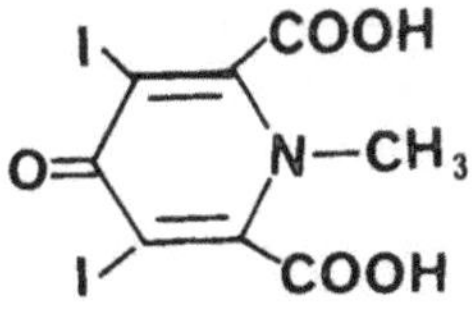

Iodoxyl

(6)

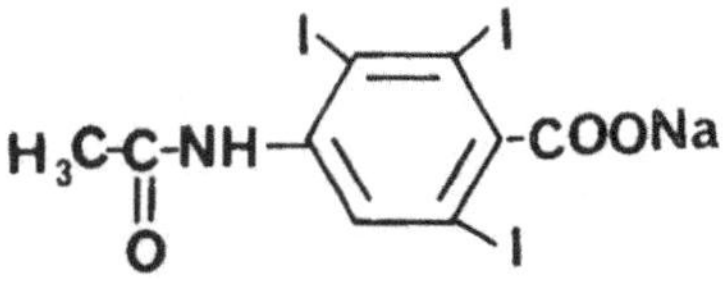

Acetriazoate (Cystokon, Salpix)

(7)

Because of the presence of at least one carboxyl group and other polar functions (e.g., CONH, C=O, or N), all these compounds are quite water soluble. The acid excretion system of the kidney is especially suitable for transporting these highly hydrophilic compounds (14).

In the gallbladder there exists a second acid excretion system, suitable for the transport of more lipophilic acids. Hence the following drugs are more suitable for cholecystography because of the introduction of a nonpolar hydrocarbon side chain or ring, resulting in a greater lipophilic character (8–11).

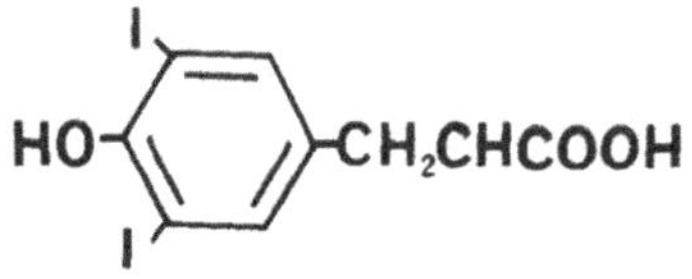

Phenidol

(8)

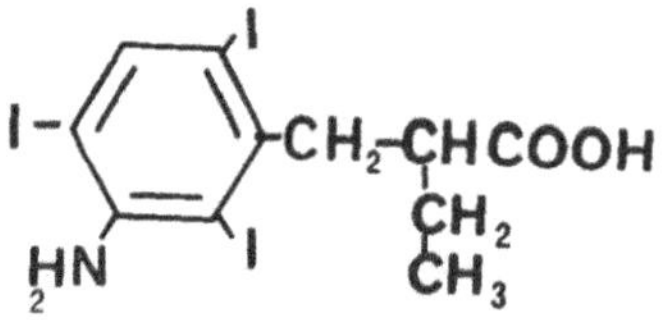

Iopanoic acid (Telepaque)

(9)

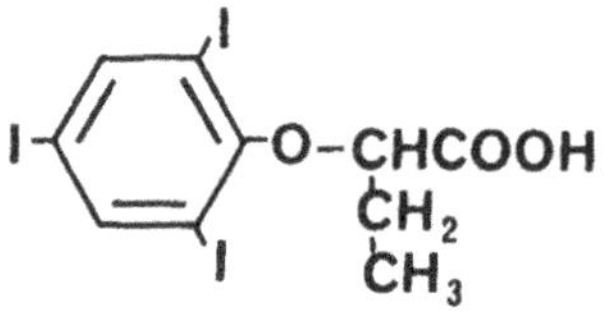

Phenobutiodyl

(10)

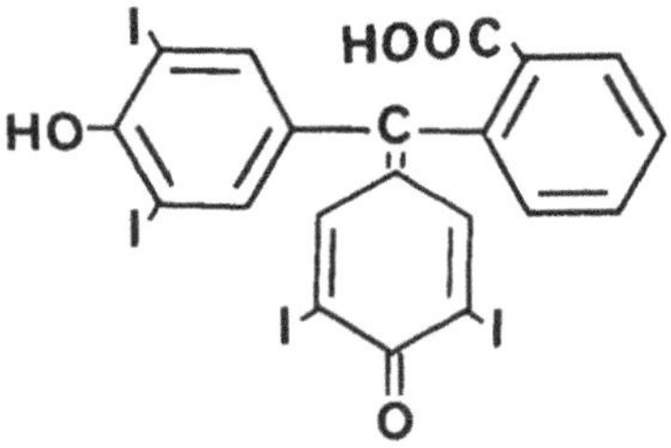

Tetraiodophenophthalein

(11)

The radiopaque agents used in cholecystography are administered orally, whereas pyelographic agents are given by intravenous injection to avoid the enterohepatic circulation and to achieve greater renal excretion.

III. MODIFICATION OF THE INTRINSIC PHARMACOLOGICAL ACTIVITY

By making structurally related compounds followed by careful testing, medicinal chemists and pharmacologists have been able to observe the change in the intrinsic activity of a parent drug not only quantitatively but also qualitatively. For example, a large number of phenothiazines or structurally related compounds have been shown to possess wide ranges of CNS activity, peripheral nervous system activity, and antibacterial or antihelmintic activity (15) (Scheme 3).

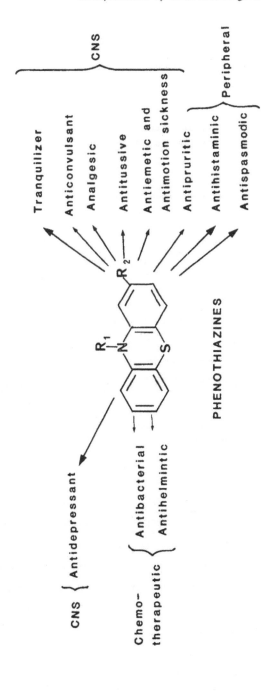

Scheme 3   Molecular modifications of phenothiazines leading to different therapeutical applications.

The spectrum of the pharmacological activity of sulfanilamide derivatives ranges from antibacteiral, antileprosy, to anticonvulsant, antidiabetic, uricosuric, and antithyroid (16) (Scheme 4).

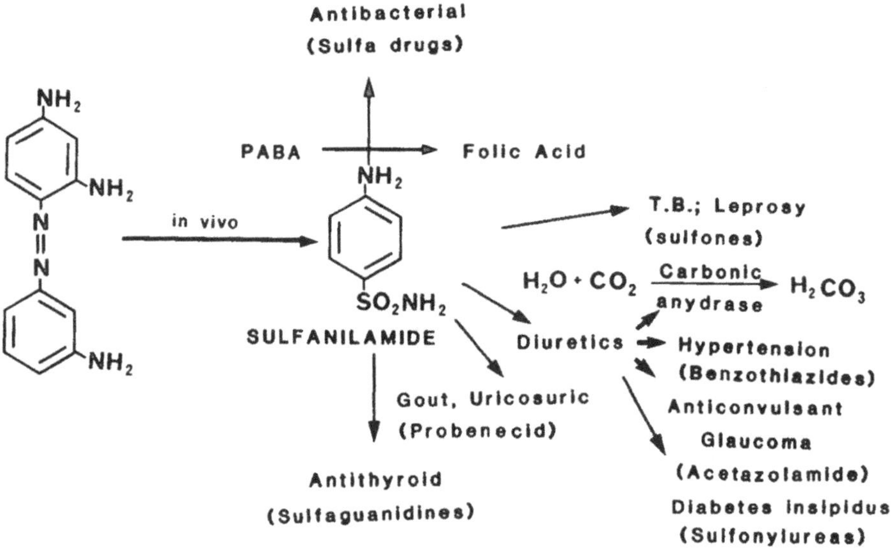

Scheme 4   Molecular modifications of sulfanilamide leading to different therapeutic agents.

Even a wider range of applications can be found among different organophosphorus compounds, ranging from anticancer agent, ophthalmic drugs, and insecticides to extremely deadly nerve gases.   While many of these organophosphates are inhibitors of acetylcholinesterase, the relative mammalian toxicity determines their usefulness either as insecticides or as therapeutic agents (Scheme 5).

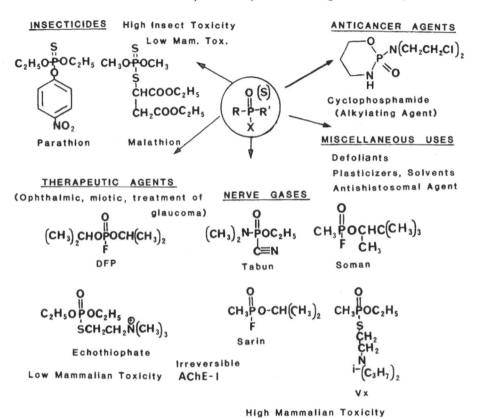

Scheme 5 Various applications of organophosphorus compounds.

In a structure-activity correlation study of a series of CNS depressant 2-imidazolidinones with a tertiary amino group on the side chain, it has been found that the CNS activities are highly dependent on the log of the octanol/water partition coefficient. Chlorpromazine, a highly potent and lipophilic compound (log P 5.35), also fits the same regression lines (17). A 1972 international double-blind clinical evaluation has reported pimozide (an imidazolidinone derivative) to be the drug of choice for long-term maintenance treatment of chronic psychotic patients, who are responsive to the antipsychotic and socializing effects of neuroleptics (18) (12). It was marketed in 1984 for the treatment of Tourette's disorder.

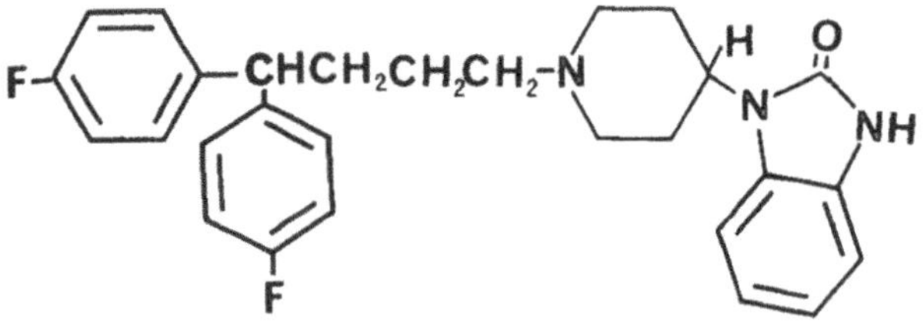

(12)

Pimozide, 1-{-1-[4,4-bis(*p*-fluorophenyl)butyl]-4-piperidyl}-2-benzimidazolinone

The incorporation of two *p*-fluorophenyl groups in the side chain and the benzene ring on the imidazolidinone greatly enhances the lipophilicity and the neuroleptic activity. The *p*-fluoro group also prevents microsomal hydroxylation at the para position of the benzene ring, and thus prolongs the duration of action

IV. MINIMIZATION OF BIODEGRADATION OF DRUGS

In the case of oral hypoglycemic agents, the *p*-methyl group of tolbutamide is oxidized very rapidly in the liver to the corresponding benzoic acid derivative. It has a half-life of about 5 hr. In chlorpropamide the *p*-chloro group is resistant to oxidation, and it also prevents the *p*-hydroxylation of the benzene ring. Consequently, chlorpropamide is not metabolically degraded to any significant degree and is excreted very slowly in unchanged form. The half-life of clorpropamide is about 36 hr, or seven times as long as that of tolbutamide (19) (Scheme 6).

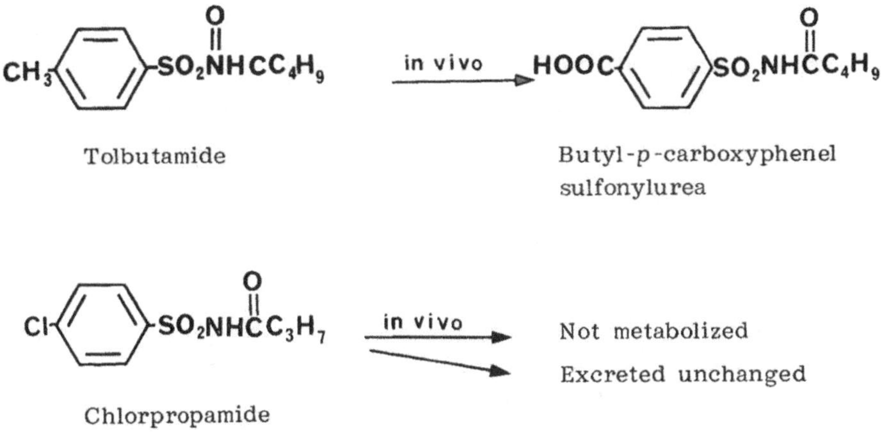

Tolbutamide   in vivo   Butyl-p-carboxyphenel sulfonylurea

Chlorpropamide   in vivo   Not metabolized / Excreted unchanged

Scheme 6   Comparison of the metabolic fate of a p-methyl group with that of a p-chloro group in sulfonylureas.

Using simple models of acetylation in the liver and excretion from the kidney, Fujita employed multiple regression analysis to correlate the rate constants of both processes with substituent constants (20). He found that for the long-acting sulfa drugs the pK_a values ideally should be between 6.0 (the urinary pH of humans) and 7.4 (the blood pH). Some examples of long-acting sulfa drugs are sulfamethoxypyridazine (pK_a 7.2), sulfamonomethoxine (pK_a 6.0), sulfadimethoxine (pK_a 6.2), and sulfamerazine (pK_a 6.9).

Sulfa drugs with pK_a values much lower than 6.0 or much higher than 7.4 tend to have a short half-life. For example, sulfisoxazole (pK_a 4.6) is used for urinary infection, and after the initial dose of 4 g, subsequent doses of 1 g should be given every 4 hr (21). Sulfanilamide itself has a pK_a value of 10.5; therefore, it has a very short duration of action.

Less absorbable sulfa drugs such as salicylazosulfapyridine, sulfaguanidine, and phthalylsulfathiazole are most often pre-scribed for ulcerative colitis (22). It should be borne in mind that a small but significant portion of the drug is absorbed; therefore, allergic reactions such as blood dyscrasias and skin rashes may also occur.

V. IMPROVEMENT OF ACCEPTABILITY BY CHEMICAL MODIFICATION

The bitter taste of many antibiotics can be abolished by pre-paring the appropriate reversible derivatives. Chloramphenicol as such has a bitter taste, but the less water soluble palmitate ester is practically tasteless (13). Hydrabamine phenoxymethyl

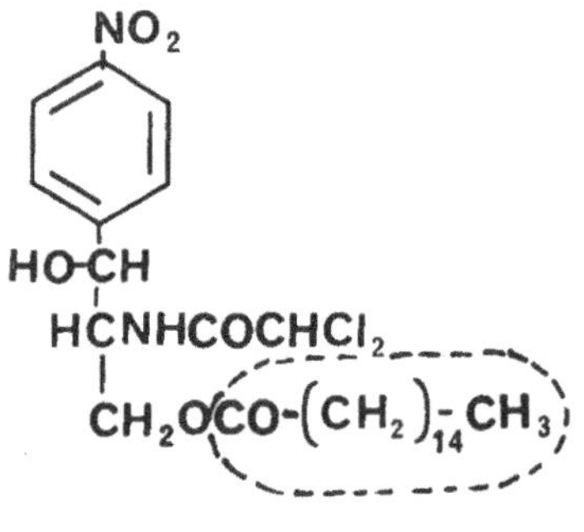

(13)

Chloramphenicol palmitate

penicillin V (Abbocillin-V, Compocillin-V) is also tasteless and has been used in oral suspension and chewable wafers (14).

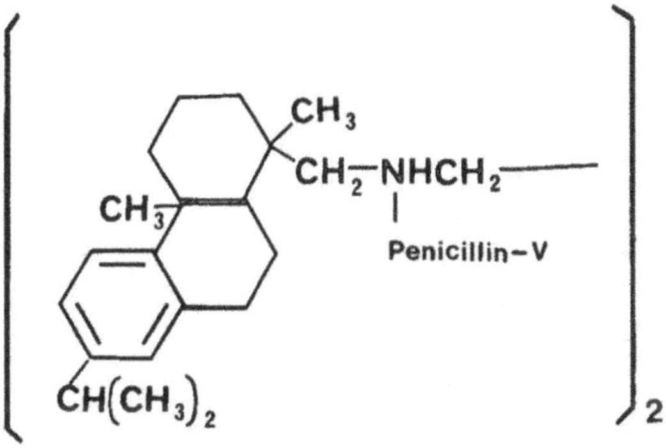

$$(\underline{14})$$

Hydrabamine penicillin V
[N,N'-bis(dehydroabietyl)ethylenediamine
bis(phenoxymethyl penicillin V)]

Clindamycin-2-palmitate and clindamycin-2-hexyldecylcarbonate HCl have recently been reported to be virtually devoid of the characteristic bitter taste of clindamycin. These two derivatives have also been shown to give blood levels equivalent to the parent drug in dogs (64) (underline15).

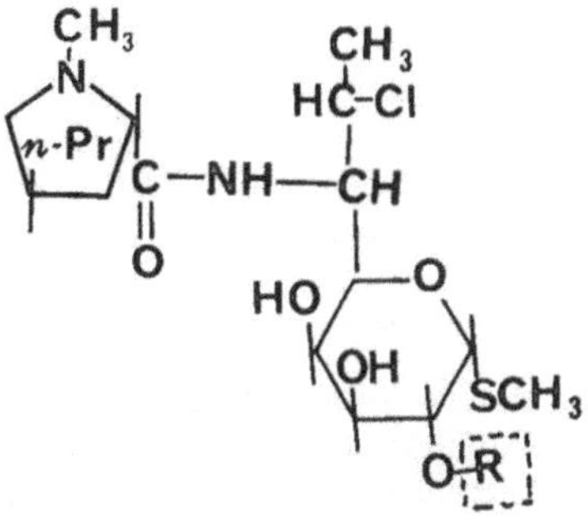

Clindamycin

$$R = H \qquad (\underline{15})$$

Reversible derivatives of clindamycin

Ethanethiol is known to be effective as an antituberculosis and antileprosy agent. Unfortunately, it is too obnoxious to be acceptable. The isophthalic acid ester, being a bland, odorless oil, has been shown to be active against experimental tuberculosis in mice and guinea pigs (23, 24). This thioester, in the presence of water, can be hydrolyzed to give the active ethanethiol (16).

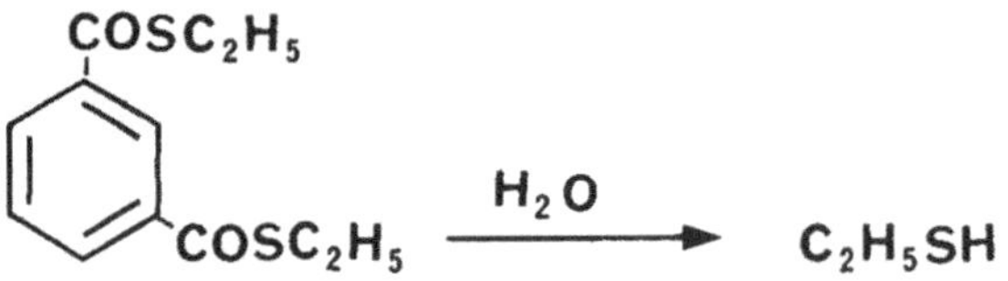

$$COSC_2H_5 \text{ ring } COSC_2H_5 \xrightarrow{H_2O} C_2H_5SH \quad (16)$$

Oily liquid   Liquid, bp 34–35°C,
                               penetrating odor

Diethyl dithiolisophthalate   Ethanethiol (ethyl mercaptan)

The para isomer, diethyl dithioterephthalate, an odorless, tasteless crystalline solid, is also active in experimental tuberculosis (24).

REFERENCES

1. Shore, P. A., Brodie, B. B., and Hogben, C. A. M. (1957). The gastric secretion of drugs: a pH partition hypothesis. *J. Pharmacol. Exp. Ther. 119*: 361.

2. Schanker, L. S., Shore, P. A., Brodie, B. B., and Hogben, C. A. M. (1957). Absorption of drugs from the stomach: I. The rat. *J. Pharmacol. Exp. Ther. 120*: 528.

3. Hogben, C. A. M., Schanker, L. S., Tocco, D. J., and Brodie, B. B. (1957). Absorption of drugs from the stomach: II. The human. *J. Pharmacol. Exp. Ther. 120*: 540.

4. Schanker, L. S., Tocco, D. J., Brodie, B. B., and Hogben, C. A. M. (1958). Absorption of drugs from the rat small intestine. *J. Pharmacol. Exp. Ther. 123*: 81.

5. Hogben, C. A. M., Tocco, D. J., Brodie, B. B., and Schanker, L. S. (1959). On the mechanism of intestinal absorption of drugs. *J. Pharmacol. Exp. Ther. 125*: 275.

6. Gibaldi, M., and Kanig, J. L. (1966). The effect of body position and pH on the gastrointestinal absorption of salicylate. *Arch. Int. Pharmacodyn. Ther. 161*: 343.

7. Lien, E. J. (1970). Physocochemical properties and gastrointestinal absorption of drugs. *Drug Intell. 4*: 7.

8. Lien, E. J. (1974). The relationship between chemical structure and drug absorption, distribution and excretion, in J. Maas, Ed., *Medicinal Chemistry IV. Proceedings of the 4th International Symposium on Medicinal Chemistry.* Elsevier, Amsterdam, pp. 319–342.

9. Lien, E. J. (1976). Structure-absorption-distribution relationships for bioactive compounds: its significance for drug design, in E. J. Ariëns, Ed., *Drug Design*, Vol. 5. Academic Press, New York, pp. 81–132.

10. Lien, E. J. (1981). Structure-activity relationships and drug disposition. *Annu. Rev. Pharmacol. Toxicol. 21*: 31.

11. T'Ang, A., and Lien, E. J. (1982). Quantitative analysis of pharmacokinetic constants as functions of physicochemical parameters. *Acta Pharm. Jugosl. 32*: 87.

12. Frisk, A. R., and Tunevall, G. (1963). In vitro activity and blood concentration studies with a series of phenoxy-alkylpenicillins. *Chemotherapia 7*: 67.

13. Daehne, W. V., Gunderson, F. E., and Lund, F., Morch, P., Peterson, H. J., Roholt, K., Tybring, L., and Godteredsen, W. O. (1970). Acyloxymethyl esters of ampicillan. *J. Med. Chem. 13*: 607.

14. Ariëns, E. J. (1966). Molecular pharmacology, a basis for drug design, in E. Jucker, Ed., *Progress in Drug Research*, Vol. 10. Birkhäuser Verlag, Basel, pp. 490–491.

15. Gordon, M., Craig, P. N., and Zirkle, C. L. (1964). Molecular modification in development of phenothiazine

drugs, in R. F. Gould, Ed., *Molecular Modification in Drug Design*, Advances in Chemistry Series 45. American Chemical Society, Washington, D.C., pp. 140–147.

16. Tishler, M. (1964). Molecular modification in modern drug research, in R. F. Gould, Ed., *Molecular Modificaiton in Drug Design*, Advances in Chemistry Series 45. American Chemical Society, Washington, D.C., pp. 1–14.

17. Lien, E. J., Hussain, M., and Golden, M. P. (1970). Structure-activity correlations for the central nervous depressant 2-imidazolidinones. *J. Med. Chem. 13*: 623.

18. Janssen, P., Brugmans, J., Dony, J., and Schuermans, V. (1972). An international double-blind clinical evaluation of pimozide. *J. Clin. Pharmacol. 12*: 26.

19. Travis, R. H., and Sayers, G. (1970). Insulin and oral hypoglycemic drugs, in L. S. Goodman and A. Gilman, Eds., *The Pharmacological Basis of Therapeutics*, 4th ed., Macmillan, New York, p. 1593.

20. Fujita, T. (1971). Substituent-effect analysis of the rates of metabolism and excretion of sulfonamide drugs, Symposium on Biological Correlation, 161st ACS National Meeting, Los Angeles.

21. Modell, W. (1972). *Drug of Choice 1972–1973*. C. V. Mosby, St. Louis, Mo., p. 610.

22. Sinkula, A. A., Morozowich, W., and Rowe, E. L. (1970). The chemical modification of clindamycin: synthesis and evaluation of selected esters. *Abstr. Papers, Amer. Pharm. Assoc. Meeting*, Washington, D.C., p. 63.

23. Davies, G. E., and Driver, G. W. (1957). The antituberculus activity of ehtyl thiolesters with particular reference to diethyl dithiolisophthalate. *Brit. J. Pharmacol. 12*: 434.

24. Davies, G. E., and Driver, G. W. (1960). Action of two ethyl thioesters against experimental tuberculosis in the guinea-pig. *Br. J. Pharmacol. 15*: 122.

Three

Quantitative Structure-Activity Relationship

I. DOSE-RESPONSE CURVE AND QUANTITATIVE MEASUREMENT OF BIOLOGICAL ACTIVITY

If it cannot be expressed with number, we do not know much about it.

Lord William Kelvin
(1824–1907)

Quantitative pharmacology is such an underdeveloped subject that it is hopeless to expect formal proof for any hypothesis and equally hopeless to expect any hypothesis to explain all the facts observed.

A. J. Clark (1937)
R. F. Furchgott (1955)
Pharm. Rev. 7: 178

A. Dose-Response Curve

There are three types of quantitative measurements by which the effect of a drug on an animal may be evaluated.

1. *Individual effect.* The individual effective dose is measured on an animal. Example: the gradual injection of a convulsant drug until convulsions begin.
2. *Graded effect.* The effect on each animal of the test group is measured, but the dose is varied. Example: the weight of an organ after injection of an anabolic agent.
3. *Quantal effect.* The all-or-none response of a group of animals is determined in order to determine the percent responding.

B.   Median Effective Dose (ED_{50}) and Median Lethal Dose (LD_{50})

A very important concept that concerns toxicity as well as
therapeutic effect is the phenomenon of *biologic variation*.   If
the dose of a drug is determined which will give an all-or-none
response—for example, the death of experimental animals in a
large population—it is found that the susceptibility of the various
individuals varies considerably, as shown in Figure 3.   This
biologic variation occurs even with inbred strains of animals.
As is to be expected, there is a much greater variation in the
heterogeneous human population.

The biologic variation in drug action is an important
reason why therapeutics must be individualized, treatment being
adjusted to the individual requirements of the patient.   It should

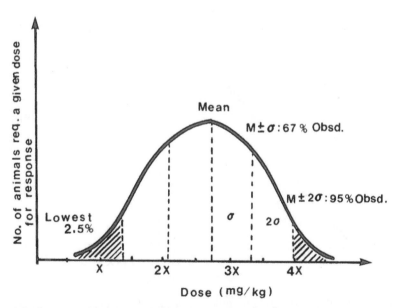

Figure 3   Biologic variation in susceptibility to drugs.

also be obvious that no sweeping generalization concerning effectiveness or safety of a drug can be made on the basis of a few clinical trials. With all the training in physicochemical, pharmacological, and clinical aspects of drug action, the pharmacist should be the best equipped person to determine the dosing and to carry out the monitoring for drug therapy.

The ED_{50} is defined as the dose effective for producing a certain sign in 50% of the animals in a group (Figure 4). The units are those of the dose (e.g., mg/kg), and the value is, of course, different for each route of administration. The ED_{50} is calculated, since it would be fortuitous that one of the doses of a series should produce the effect in exactly half of the animals, and the ED_{50} could not occur even fortuituously if the group contained an odd number of animals. When the all-or-

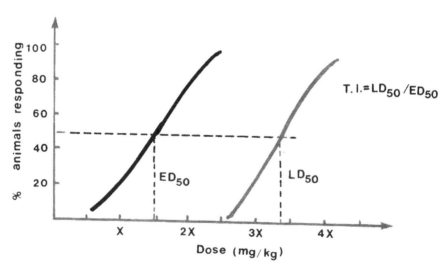

Figure 4 Illustration of concept of median effective dose (ED_{50}), median lethal dose (LD_{50}), and therapeutic index (LD_{50}/ED_{50}). In a similar way, safety factor can be defined as LD_1/ED_{99}.

none response, also called the quantal response, is death, the ED_{50} becomes the LD_{50}, or the lethal dose for 50% of the animals.

Sometimes the ED_{75}, the ED_{99}, and the ED_{10}, for example, are desired in order to know a dose affecting most of the animals, a nearly minimally effective dose, and so on. These doses can be calculated conveniently by a graphical method.

C. Estimation of the ED_{50} and Its Error by Means of Logarithmic–Probit Graph Paper

The graphic procedure of Litchfied and Fertig (1) has been simplified and extended by Miller and Tainter (2,3). This method is probably the most convenient and useful for calculating any ED value. A special coordinate paper is used, logarithmic-probit paper, which has two decades of a logarithmic scale as the abscissa and a scale of probits as the left ordinate. Preferably, the ordinate is marked in a scale of percent corresponding to the probit scale (this percent scale is nonlinear; see Figure 5). If the percent scale is absent, the probit corresponding to a percent value must be found in a table of probits (4, 5) and plotted on semilog paper (Figure 6).

The theory of probits is discussed in many books on statistics. Essentially, the probit transformation effects a stretching of the ends of the sigmoidal curve, formed by plotting the quantal response versus the logarithm of dose, so that the sigmoid curve becomes a straight line.

In the following example taken from the book of Turner (3), the total alkaloids of belladonna bellafoline is given to mice orally to determine the acute LD_{50}. Groups of 10 mice, 5 male and 5 female, are used. The volume of all doses given to

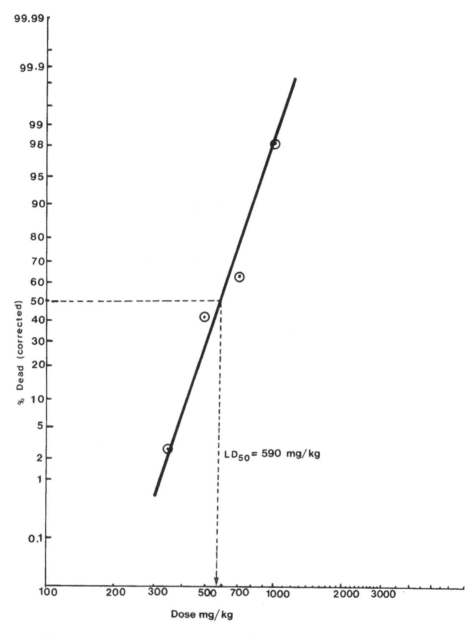

Figure 5　Dose-response curve of the acute lethal toxicity of bellafoline plotted on probability-2-log cycle paper.

the mice is 0.025 ml of solution per gram of body weight. A
solution of the drug in 2% malic acid is prepared, containing
40 mg/ml, which is diluted to give solutions of 28, 20, 14, and
10 mg/ml (the ratio of a dose to its antecedent is 1:2). The time
of death of the killed animal is recorded. The results are
shown in Table 5.

   This is a "stepping-down" procedure. If one starts from
the lower end of the dose-response curve, one can use a "step-
up" procedure. Most accurate results can be obtained if one

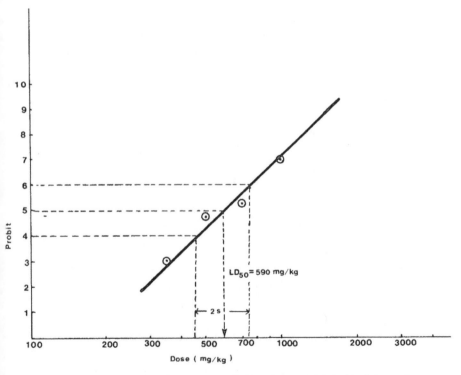

Figure 6 Dose-response curve of the acute lethal toxicity of
bellafoline using probit units on two-cycle semilogarithmic paper.

TABLE 5 Acute Oral Toxicity of Bellafoline in Mice Computed by the Miller-Tainter Method

Group	Dose (mg/kg)	Number of mice died	Number of mice survived	% Dead	Corr. %[a]	Probit
1	1000	10	0	100	97.5	6.96
2	700	6	4	60	60	5.25
3	500	4	6	40	40	4.25
4	350	0	10	0	2.5	3.04
5	250	0	10	0	—	—

[a]Correction formulas (n is the number of animals in a group): For the 0% dead, corrected % = 100(0.25/n); for 100% dead, corrected % = (100n − 25)/n.

knows the approximate ED_{50} or LD_{50} value of the test compound and tries to zero in the median value by find adjustment of the doses, just as in the case of target shooting.

Since all of the mice in group 4 (Table 5) survived, group 5 is neglected. The percents for the first and last groups are corrected before plotting, according to the formulas given in Table 5. The probit values are read from a table of probits (4), and they or the corrected percent values are plotted against the logarithms of the dose. The dose corresponding to 50%, or a probit of 5, is found to be 590 mg/kg (see Figures 5 and 6).

To compute the standard deviation of the mean, D, the doses for probits 4 and 6 were read from Fig. 6. They are found to be 455 and 780 mg/kg, respectively. Their difference being defined as 2S, and from them D, is computed as follows:

$$D = \frac{2S}{\sqrt{2 \times 2n}} = \frac{780 - 455}{\sqrt{40}} = 51.4 \text{ mg/kg}$$

Thus the LD_{50} of bellafoline is reported as 590 ± 51 mg/kg.

D. Affinity Constants pA_2 and pD_2'

These two expressions of biological activities are commonly used. They are useful for comparison of dose-response curves of congeners according to the following definitions (6, 7):

$$pA_2 = -\log[B] + \log\left\{\frac{[A_2]}{[A_1]} - 1\right\}$$

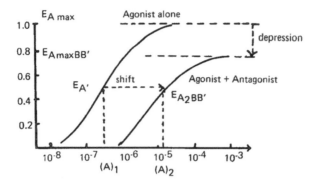

Figure 7 Dose-response curves showing the parallel shift and the depression.

when $[A_2]/[A_1] = 2$, $pA_2 = -\log[b]$. Thus pA_2 is calculated from the shift of the dose-response curve (Figure 7). It is known as the affinity constant (some authors call it log K). It is the negative log (molar concentration of the antagonist) B, which requires doubling of the agonist concentration [A] to achieve the same effect.

$$pD'_2 = -\log[B] - \log\left(\frac{E_{A\ max}}{E_{A\ max} - E_{A\ max}BB'} - 1\right)$$

pD'_2 is calculated from the depression of the dose-response curve.

Competitive antagonism leads to a parallel shift of the dose-response curve, while noncompetitive antagonism may lead to a nonparallel shift or depression of the curve. The noncompetitive antagonism can be further differentiated into metaffinoid antagonism and metactoid antagonism.

1. *Metaffinoid antagonism.* The change in the receptor of
   the antagonist results in a change in receptor of the
   agonist. In other words, in the presence of the antagonist, the affinity of the agonist to the receptor is
   decreased. A metaffinoid system is similar to a "pure
   K system" in enzymology.

2. *Metactoid antagonism.* The change in the receptor of
   antagonist leads to an interference with the ability of
   the receptor-effector system of the agonist to respond
   to the interaction between the agonist and its receptor.
   This is like the "pure V_{max} system" in enzymology.

E. Intrinsic Activity and pD_2

The intrinsic activity of an agonist is determined according to
the definition $\alpha^E = E_{A_{max}}/E_{max}$, where $E_{A_{max}}$ is the maximal
effect that can be obtained with the test compound, and E_{max} is
is the maximal effect of a reference compound on the same test
organ. The pD_2 value of an agonist is the $-\log A$, which pro-
duces 50% of the maximal effect of the drug on the receptor-
effector system (6).

II. KINETICS VERSUS THERMODYNAMICS

A rate constant (kinetics) indicates *how fast* a reaction or
process can proceed; it does not indicate how complete the
reaction will go:

$$A \underset{k_{-1}}{\overset{k}{\rightleftharpoons}} B \qquad k = Ae^{-E_a/RT}$$

Forward rate = $k[A]$ $\qquad \log k = \log A - \dfrac{E_a}{2.303RT}$

Backward rate = $k_{-1}[B]$

An equilibrium constant (thermodynamic), on the other hand, describes *how far* a reaction or process can proceed— whether it takes a few seconds or many years: Under equilibrium conditions one has

$$k[A] = k_{-1}[B]$$

$$K = \frac{k}{k_{-1}} = \frac{[B]}{[A]}$$

K is related to the standard free energy change by

$$\Delta G° = - RT \ln K$$
$$= - 2.303RT \log K$$

In the partitioning process:

$$\Delta G° = -2.303RT \log P$$

$$P = \frac{[drug]_o}{[drug]_{aq}}$$

The rate constant k is related to the equilibrium constant $K\ddagger$ (the double dagger refers to the transition state) for the process:

$$\text{reactants} \; \underset{}{\overset{k}{\rightleftharpoons}} \; \text{transition state}$$

$$k = \frac{RT}{Nh} \, K\ddagger$$

R = gas constant

T = absolute temperature

N = Avogadro's number

h = Planck's constant

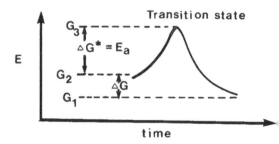

Since $\Delta G^{\ddagger} = -RT \ln K^{\ddagger}$, we have

$$\log k = \frac{-\Delta G^{\ddagger}}{2.303RT} + \log \frac{RT}{Nh}$$

$$= \frac{-E_a}{2.303RT} + \log A$$

$$\Delta G^{\ddagger} = a\Delta G^{\circ} + RT \left(\log \frac{RT}{Nh} - b \right)$$

Therefore, log k and log K are *linearly related* to the *free energy* change of the process.

The question of how molecular modification affects K or k of a reaction was answered quantitatively by Hammett around 1940 (8). The Hammett equation is defined as (17)

$$\log \frac{K}{K_o} = \rho\sigma \quad \text{or} \quad \log \frac{k}{k_o} = \rho\sigma$$

$$\sigma = \log \frac{K_{XPhCOOH}}{K_{PhCOOH}} \qquad \rho = \text{reaction constant}$$

$$\qquad\qquad\qquad\qquad\qquad \sigma = \text{substituent constant}$$

$$\sigma = PK_{a_{BA}} - PK_{a_{XBA}}$$

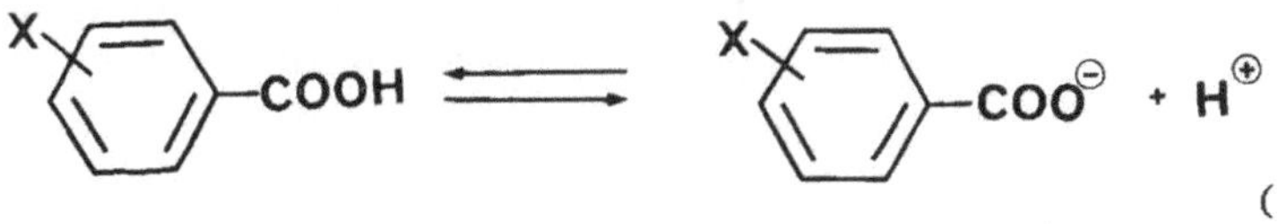

$$(17)$$

$$K = \frac{[XPhCOO^-][H^+]}{[XPhCOOH]}$$

Electron-withdrawing groups such as NO_2, CF_3, CN, Cl, and Br will weaken the COO—H bond and stabilize the $XPhCOO^-$ ion, thus increasing K, and giving lower pK_a values and stronger acids. Electron-donating groups such as CH_3,

NH_2, and $N(CH_3)_2$ will strengthen the COO—H bond and facilitate the recombination of $XPhCOO^-$ and H^+, thus giving higher pK_a values and weaker acids.

Given that many types of biological activities are goverend by both pharmacokinetic and pharmacodynamic processes, one may wonder how the linear free energy relationship (LFER) can be used in quantitative structure-activity relationship analysis for series of congeners (9).

Since both log k and log K are linear functions of free energy change, this should not impose a problem. In other words, they can be used in the correlation study. This is shown by the following equations (10):

$$\left. \begin{array}{l} \log K = \dfrac{-\Delta G^\circ}{2.303RT} \\[2ex] \log P = \dfrac{-\Delta G^\circ}{2.303RT} \end{array} \right\}$$ for equilibrium process and phase distribution, respectively

$$\log k = \log \dfrac{RT}{Nh} \dfrac{-\Delta G^\ddagger}{2.303RT}$$ for a specific rate constant going through a transition state

$$\left. \begin{array}{l} \Delta G^B = a\Delta G^A + b \\[2ex] \log P_B = c \log P_A + d \end{array} \right\}$$ for two different equilibrium systems, A and B

In the Hammett equations σ is a measure of the electron density on the benzene ring when a substituent group is attached. Some of the most common electron-donating and electron-withdrawing groups are given with their σ values at either the meta or the para position (Table 6). The reaction constant ρ, on the other hand, is a measure of the sensitivity of the reaction to the substituent effect (11) (Table 7).

TABLE 6 Some Common Electron-Donating and Electron-Withdrawing Groups With Opposite Signs in Sigma Constants

Electron-donating group	σ_m	σ_p	Electron-withdrawing group	σ_m	σ_p
$-CH_3$	−0.07	−0.17	$-NO_2$	+0.71	+0.78
$-C_2H_5$	−0.04	−0.15	$-F$	+0.34	+0.06
$-NH_2$	−0.16	−0.66	$-Cl$	+0.37	+0.23
$-N(CH_3)_2$	−0.15	−0.83	$-CN$	+0.68	+0.63
$-i-C_3H_7$	−0.07	−0.13	$-SO_2NH_2$	+0.46	+0.57

The Hammett equation applies only to aromatic systems; for aliphatic systems, Taft's polar substituent constant was derived from the acid- and base-catalyzed hydrolysis of esters (11) (Scheme 7).

TABLE 7 Relative Reaction Constants of Different Chemical · Processes

Reaction	ρ
Ionization of benzoic acids, water 25°C (eq.)	1.000
Ionization of benzoic acids, ethanol, 25°C	1.957
Ionization of phenols, water, 25°C	2.113
Saponification of methylbenzoates, 60% acetone, 0°C (rate)	2.460
Benzoylation of aromatic amines (benzene)	−2.781
cis-trans isomerization of azobenzenes	−0.610
Side-chain bromination of acetophenones	0.417

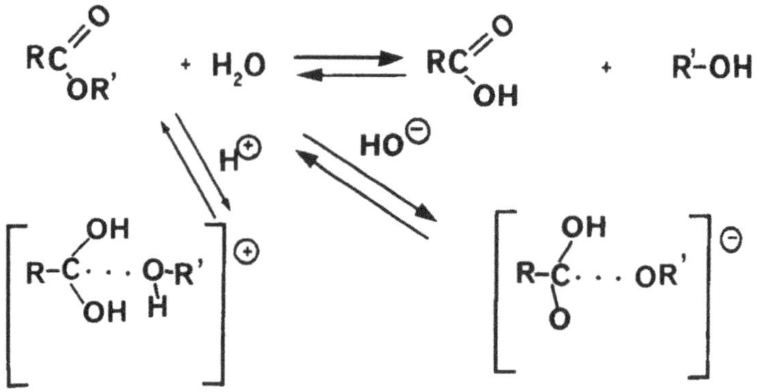

(A) H^+-catalyzed
transition state

(B) OH^--catalyzed
transition state

Electronic effect of R
minimum

Polar effect + steric effect

$$E_s = \log \left(\frac{k}{k_o}\right)_A \qquad \log \left(\frac{k}{k_o}\right)_B = \rho^* \sigma^* + sE_s$$

k_o = rate constant of
the acetate (R = CH_3)

$$\sigma^* = \frac{1}{2.48} \left[\log \left(\frac{k}{k_o}\right)_B - \log \left(\frac{k}{k_o}\right)_A \right]$$

Scheme 7

Since the intermediate (A) has only two extra H's compared to (B), the steric effect should be virtually the same (11, 12).

The polar and the steric substituent constants of some functional groups are shown in Table 8.

The rate of saponification of substituted benzoic acid esters has been shown by Hancock and Falls to be well correlated with the σ of the ring substituent and the σ^* and E_s^c of the aliphatic ester chain (13) (18).

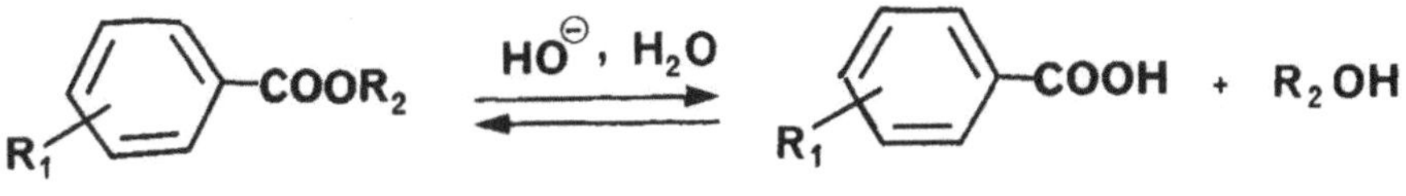

$$\log\ k\ (M^{-1}\ m^{-1}) = 0.174 + 2.22\sigma_{(R_1)} + 1.53\sigma^*_{(R_2)} + 0.668E^c_{s_{(R_2)}} \quad (\underline{18})$$

n	r	s
35	0.996	0.108

$E^c_s = E_s + 0.306(n - 3)$ steric constant corrected for hyperconjugation effect; n = number of H hyperconjugated

TABLE 8 Polar and Steric Substituent Constants of Some Functional Groups

Substituent (R in R—COOR')	σ^*	E_S
Cl_3C-	2.65	−2.06
Cl_2CH-	1.94	−1.54
$ClCH_2-$	1.05	−0.24
$H-$	0.49	+1.24
$C_6H_6CH_2-$	0.22	−0.38
CH_3-	0.00	0.00
C_2H_5-	−0.10	−0.07
$n\text{-}C_3H_7-$	−0.12	−0.36
$i\text{-}C_3H_7-$	−0.19	−0.47
$n\text{-}C_4H_9-$	−0.13	−0.39
$s\text{-}C_4H_9-$	−0.21	−1.13
$t\text{-}C_4H_9-$	−0.30	−1.54

TABLE 9 Kinetic Constants and Equations for Zero-, First-, and Second-Order Reactions

Order	Initial concentration	Differential equation	Integrated equation ($x = 0$ when $t = 0$)	Equation for linear plot	Slope of linear plot (t as abscissa)
Zero	a	$\dfrac{dx}{dt} = k$	$kt = x$	$x = kt$	k
		Half-life $t_{1/2} = \dfrac{(1/2)a}{k} = \dfrac{a}{2k}$			
First	a	$\dfrac{dx}{dt} = k(a - x)$	$kt = \ln \dfrac{a}{a - x}$	$\log(a - x) = \log(a) - \dfrac{kt}{2.3}$	$\dfrac{-k}{2.3}$
		Half-life $t_{1/2}\ (\tau) = \dfrac{2.3 \log 2}{k} = \dfrac{0.693}{k}$		For two parallel processes	
				$\dfrac{1}{t} = \dfrac{1}{t_1} + \dfrac{1}{t_2}$ or $t = \dfrac{t_1 t_2}{t_1 + t_2}$	
Second	a = b	$\dfrac{dx}{dt} = k(a - x)^2$	$kt = \dfrac{x}{a(a - x)}$	$\dfrac{1}{a - x} = \dfrac{1}{a} + kt$	k
		Half-life $t_{1/2} = \dfrac{1}{ak}$			
	a ≠ b	$\dfrac{dx}{dt} = k(a - x)(b - X)$	$kt = \dfrac{1}{a - b} \ln \dfrac{b(a - x)}{a(b - x)}$	$\log \dfrac{a - x}{b - x} = \log \dfrac{a}{b} + \dfrac{kt\,(a - b)}{2.3}$	$\dfrac{k(a - b)}{2.3}$

58

It is interesting to note that the pK_a values of various amines can be correlated with σ^* values. When the number of hydrogens of the protonated amine (n_H) is included, all three types of amines can be included in a single equation (14):

		n	r	s
1° amines:	$pK_a = -3.201\sigma^* + 13.214$	27	0.989	0.257
2° amines:	$pK_a = -2.931\sigma^* + 11.631$	22	0.986	0.316
3° amines:	$pK_a = -3.243\sigma^* + 9.558$	43	0.995	0.165
All amines:	$pK_a = -3.140\sigma^* + 1.816n_H$			
	$+ 7.817$	92	0.985	0.299

The equations given in Table 9 can be applied to either chemical reactions or pharmacokinetic processes (15).

III. STRUCTURE-ACIDITY-SOLUBILITY RELATIONSHIP

The following section gives a discussion of the acid-base and solubility properties of drug molecules based on chemical structure and functional group analysis. A thorough knowledge in this topic is needed in order to estimate the relative acid-base strength of a new drug and its relative solubilities in common solvents, including water at different pH values, and to be able to predict or explain incompatibilities due to changes in solvents, pH, or different forms of drugs used in the formulation.

A. Definitions of Acids and Bases

1. *Brönsted-Lowry theory.* This theory is based on the exchange of a proton. According to this theory, an acid is a proton donor, whether it be charged or uncharged (e.g., HCl, H_2SO_4, H_3O^+). A base is a proton acceptor, be it a charged or an uncharged species (e.g., $NaOH$, OH^-, CO_3^{2-}).

2. *Lewis electronic theory.* This theory is based on the exchange or sharing of an unshared electron pair. A Lewis acid is a molecule or ion that accepts a lone pair of electrons to form a coordinate covalent bond [e.g., H^+, H_3O^+, $B(CH_3)_3$]. A Lewis base is a species capable of donating lone-pair electrons (e.g., $:NH_3$, $R_3N:$, OH^-).

3. *Conjugate pair of acid-base.* $pK_w = pK_a + pK_b = 14$ for a conjugate pair in H_2O, 25°C.

$$HCl + H_2O \rightleftharpoons H_3O^+ + Cl^-$$

Acid 1   Base 2   Acid 2   Base 1

Conjugate pair

Conjugate pair

H_2O^+ is the conjugate acid of H_2O

Cl^- is the conjugate base of HCl

4. *Potential acidic groups (A—H).*

$\geq C-H$ $-O-H$ $-S-H$ $\geq N-H$ and conjugate acids of basic groups (BH^+; see below)

5. *Potential basic groups (B:).*

$\overset{\diagdown}{\diagup} O:$ $\overset{\diagdown}{\diagup} S:$ $-\overset{\diagup}{\diagdown} N:$ and conjugate bases of acid groups (A^-; see above)

B. Factors Affecting Acidity or Basicity of Molecules

Extrinsic factors: solvent polarity, ionic strength, temperature, etc.

Intrinsic factors: molecular structure, functional group, electronic (inductive and resonance), steric, and H bonding.

1. General Expressions and Rules

$HA \rightleftharpoons H^+ + A^-$ e.g., $X-CH_2COOH \rightleftharpoons H^+ + X-CH_2COO^-$

$B:H^+ \rightleftharpoons B: + H^+$ e.g., $R_3N:H^+ \rightleftharpoons R_3N: + H^+$

2. Inductive Effect

a. Electron-withdrawing (electronegative) groups such as halogens will weaken the H—A bond and stabilize the anion, and therefore enhance the acidity (lower pK_a). Electron-donating groups (e.g., alkyl, amino) will, on the other hand, strengthen the H—A bond or enhance the electron density around B, and increase proton affinity. This will augment the basicity (lower pK_b and higher pK_a).

b. Electronegativity and inductive effect. Electronegativity is a measure of the atom's ability to attract an electron. For example, the trend of the electronegativities of various elements in the periodic table is shown as the following:

```
H    2.1                              ⟶ Most electronegative
Li   1.0    C 2.5  N 3.0  O 3.5  F 4.0
Na   0.9    Si 1.8 P 2.1  S 2.5  Cl 3.0
K    0.8                        Br 2.8
                                I  2.4
Cs   0.7
```

Least electronegative

    c. The greater the difference in the electronegativity be-
tween X and H, the stronger will be the acid H—X (higher ionic
character)—thus the order of acidity

$$HCl > HBr > HI > HOH > CH_3OH > CH_4$$

    d. *Exceptions to the rules.* Due to the significant H-bond-
ing effects of the highly electronegative small atoms, the acidity
of HF < HCl and HOH < HSH. In these special cases, the smaller elements (F and O) have greater H-bonding capability than
the larger ones (Cl and S, respectively).

    e. Inductive effect increases with increasing number of
electronegative substituents (X).

Acidity: $Cl_3C—COOH > Cl_2CHCOOH > ClCH_2COOH$

pK_a: 0.65   <   1.3   <   2.86

    f. Inductive effect decreases with increasing distance
between the electronegative group X and the dissociable hydrogen.

Acidity: $ClCH_2COOH > Cl(CH_2)_2COOH > Cl(CH_2)_3COOH$

pK_a: 2.86   <   4.06   <   4.52

3. *Resonance (Electromeric Effect, Delocalization of Electrons)*

a. Functional groups such as p-NO_2 or p-C≡N will enhance the acidity of benzoic acid or phenol by weakening the O—H bond and also by stabilizing the anion, due to contributions from the following resonance structures (Scheme 8):

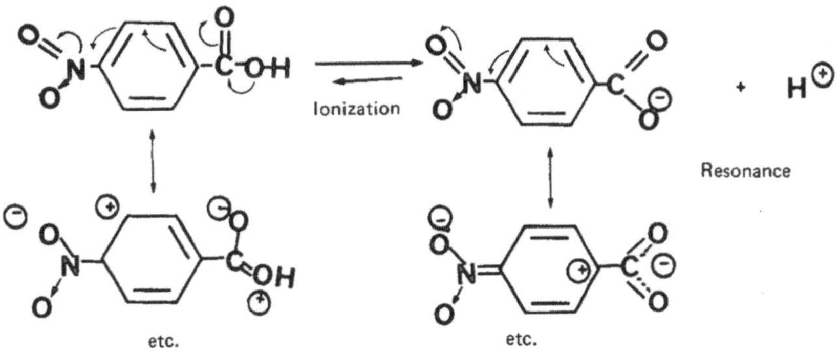

Scheme 8

Consequently, the following order of acidity exists (Scheme 9):

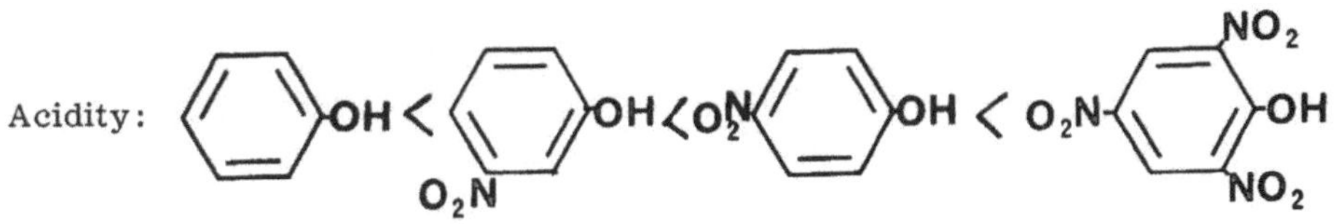

Phenol　　　　Picric acid

pK_a: 10.0 > 8.28 > 7.15 > 0.38

Scheme 9

Note: Resonance can take place only at the o, or p, position but not at the m position, while the order of inductive effect is o, m, p. The strong acidity of picric acid is due to both inductive and resonance effects.

Exercise: Write all the resonance structures of picrate ion.

b. Functional groups such as $-\ddot{N}H_2$ and $-\ddot{N}(CH_3)_2$ will donate electrons by resonance, therefore strengthen the O—H bond and enhance the recombination of the anion with the proton; consequently, weaker acids will be obtained (Scheme 10).

Acidity: H_2N⟨benzene ring⟩COOH $<$ ⟨benzene ring⟩COOH

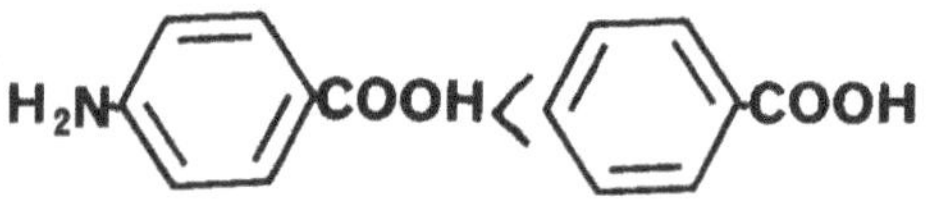

pK_a: 4.92 $>$ 4.19

Scheme 10

c. Now consider both the inductive and resonance effects. Arrange the order of acid strength of the following monosubstituted acetic acids: $X-CH_2COOH$.

$X = -NO_2 > NH_3^+ > -C{\equiv}N > F- > -COOH > Cl- > Br- >$

$-CF_3 > I- > CH_3COO- > CH_3O- > -SH > -OH >$

$-NH_2 > -Ph > CH_2{=}CH- > H- > -CH_3 > -C_2H_5 > -COO^-$

Note: Although HO— is more electronegative than CH_3O-, the acidity of the HO— substituted acetic acid may be less due to H bonding.

4. *Hydrogen Bonding (H Bonding, Hydrogen Bridge)*

    X—H. . . .X   X—H. . . .Y

where X and Y are electronegative atoms such as O, N, F, etc.

a. When the acidic H atom is involved in a H bond, it is stabilized (more energy is needed for its removal) and the acidity is lower than expected (e.g., HF, HOH, *o*-nitrophenol, *o*-halophenol).

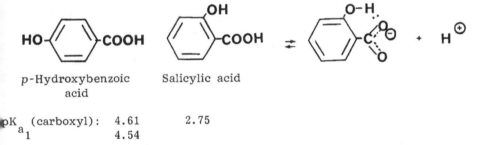

pK$_a$ 7.23   7.14

Scheme 11

Note: From the inductive effect alone one would expect the orthoisomer to be a stronger acid.

b. When the base form is stabilized by intramolecular H bonding, acid strength is enhanced.

p-Hydroxybenzoic   Salicylic acid
acid

pK$_{a_1}$ (carboxyl):   4.61   2.75
                          4.54

Scheme 12

5. *Steric Effect*

a. Since the proton is small, proton acid-base reactions are not particularly sensitive to steric compression. Nevertheless, in a highly branched (shielded) acid the anion will be less stable because of a lesser degree of solvation, and the pK_a value will be higher (a weaker acid) (16).

CH₃COOH

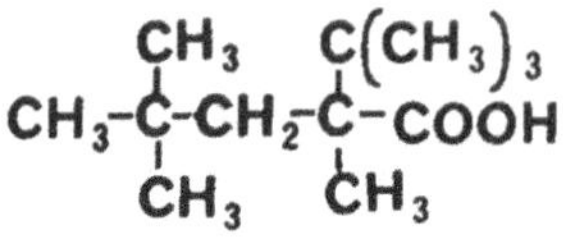

Acetic acid

pK_a = 5.57 (in 50% MeOH)

$$CH_3-\underset{\underset{CH_3}{|}}{\overset{\overset{CH_3}{|}}{C}}-CH_2-\underset{\underset{CH_3}{|}}{\overset{\overset{C(CH_3)_3}{|}}{C}}-COOH$$

pK_a = 6.96 (in 50% MeOH)

Scheme 13

TABLE 10 Effect of the 2,6-di-alkyl Groups on the pK_a Value of Pyridine

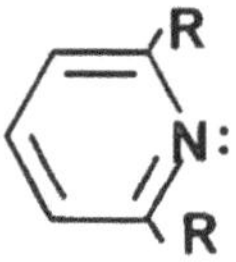

R = H, pyridine pK_a = 4.38

R	$2-R-C_5H_4N$	$2,6-R_2C_5H_3N$
Me	5.05	5.77
i-Propyl	4.82	5.34
t-Butyl	4.68	3.58

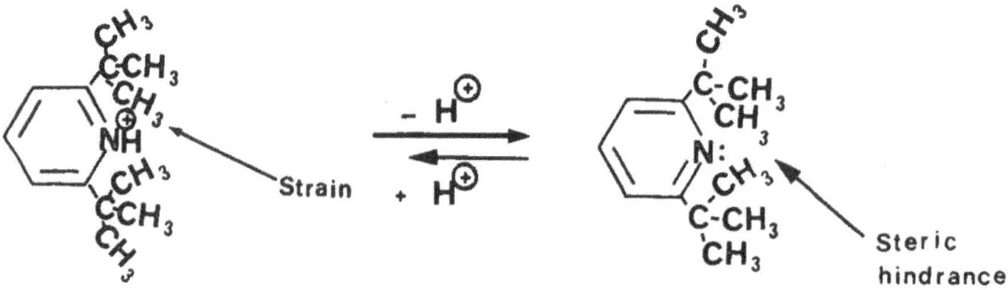

pK_a = 3.58
(a stronger acid than expected)

Strain

$- H^\oplus$
$+ H^\oplus$

Steric hindrance

(a weaker base than expected)

b. Steric hindrance may also prevent protonation, thus lowering the basicity (higher pK_b and lower pK_a). For example, 2,6-di-t-butylpyridine is weaker than expected by more than 1 pK_a unit (17) (Table 10).

Some of the prototypes of acid-base functional groups and specific pharmaceutical examples are given in Table 11 (18—21). The classification of the acids and bases is in accordance with that of Davidson (22), except that either "acidic" or "basic" is used instead of the "intermediate" class (Table 12).

C. Solubilities of Drugs

1. Definition and General Considerations

Solubility is defined in quantitative terms as the concentration of solute in a saturated solution at a certain temperature. Usually at room temperature (25°C unless specified otherwise).

$$\text{solid (solute)} \underset{k_{-1}}{\overset{k}{\rightleftharpoons}} \text{solute in solution saturated (in equilibrium)}$$

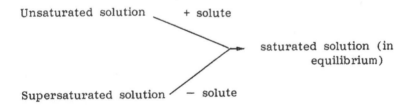

Unsaturated solution + solute

saturated solution (in equilibrium)

Supersaturated solution − solute

2. Factors Affecting Solubility

a. The solute, molecular structure, function group, polymorphism, etc.

b. The solvent, polarity, H bonding, ionic strength, cosolvent, solubilizing agent, etc.

TABLE 11 Prototypes of Acid-Base Functional Groups and Some Pharmaceutical Examples

Functional Group	Category	pK_a	Comments	Examples
Phosphate esters				
$R-O-P(OH)_2$	Acidic	pK_1 ca. 2 pK_2 ca. 7	Like H_3PO_4	Dexamethasone sodium phosphate
Sulfonic acids				
$R-SO_2OH$	Strongly acidic	<1	Like H_2SO_4	Suramin sodium, sulfobromophthalein
Sulfamic acids				
R_2NSO_2OH	Strongly acidic	<1	Like H_2SO_4 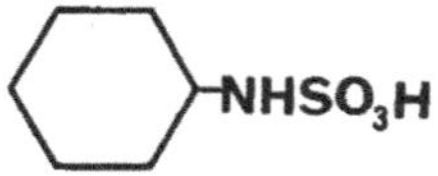	Cyclamic acid, cyclamate
Sulfate esters				
$R-OSO_2OH$	Strongly acidic	<<1	Like H_2SO_4	Heparin
Phenolic OH				
$Ar-OH$	Weakly acidic	10.0	Resonance stabilization of the anion	Phenol C_6H_5-OH
				4-Hexylresorcinol
		9.85		The phenolic OH (3-position) in morphine

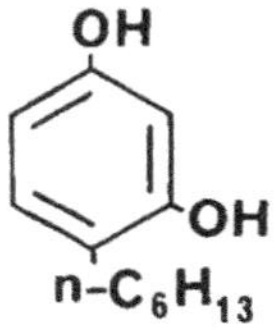

Class	Character	pKa	Remarks	Examples

Aliphatic amines
$R-NH_2$ (1°)
R_2NH (2°)
R_3N (3°)
	Basic	9.94
		
		8.2
			($\bar{c}a$. 10 usually influenced by inductive and steric effects)	$Ph-CH_2CH-NH_2$
				          $\mid$
				          CH_3
				Amphetamine
				3° amine
				Codeine
				Diphenhydramine
				$Ph-CHOCH_2CH_2N\begin{smallmatrix}CH_3\\CH_3\end{smallmatrix}$ → 3° amine
				Ph

$\underset{\displaystyle H_2N-\overset{NH}{\overset{\|}{C}}-NH_2}{}$ Guanidine	Strong base	12	Resonance stabilization of the cation	Guanethidine, streptomycin
R_4N-OH	Quaternary ammonium hydroxides	>15	Extremely strong base (like NaOH); however, the quaternary ammonium halides are neutral (like NaCl)	Benzalkonium choline, acetylcholine

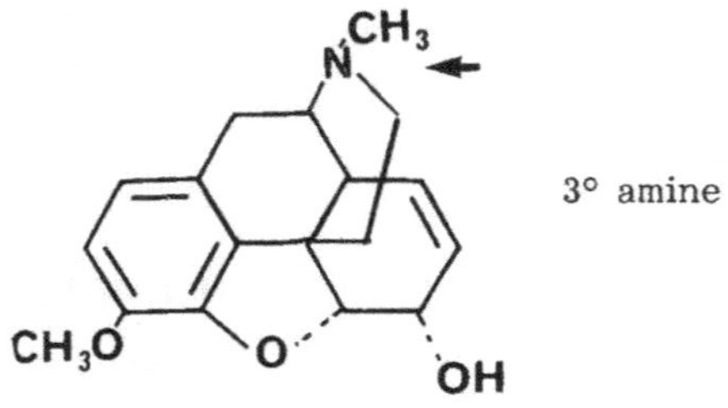

TABLE 11 (Continued).

Functional Group	Category	pKa	Comments	Examples

Basic — pKa 7.45

Comments: Lowered by the inductive effect of CONH and the steric effect of the ring and C_3H_7

Examples: Clindamycin

3° amine

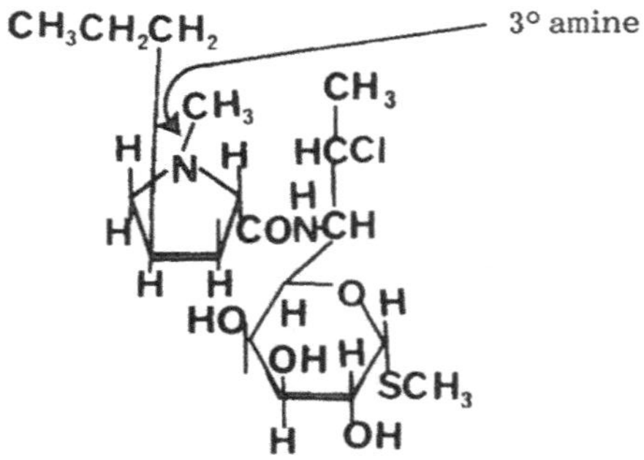

Look up the structure of lincomycin and predict its pKa.

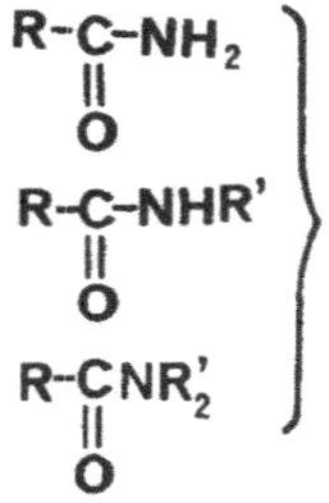

$R-\overset{\parallel}{\underset{O}{C}}-NH_2$

$R-\overset{\parallel}{\underset{O}{C}}-NHR'$ } Amides   — Neutral

$R-\overset{\parallel}{\underset{O}{C}}NR'_2$

Too weak to be measured in water

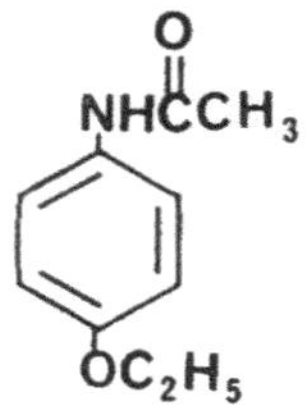

Phenacetin:

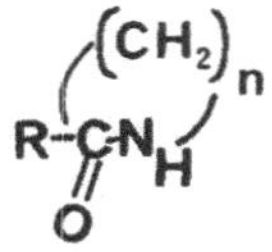

$R-\overset{(CH_2)_n}{\underset{\underset{O}{\parallel}}{C}}\overset{}{N}H$ Lactams

the β-lactam ring in penicillins and cephalosporins

70

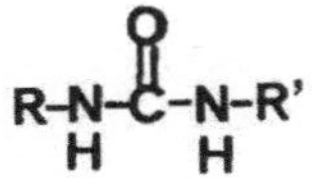

$R-\underset{H}{N}-\overset{\overset{\displaystyle O}{\|}}{C}-\underset{H}{N}-R'$ Urea derivatives

Neutral

$\underset{\text{Urea}}{H_2N-\overset{\overset{\displaystyle O}{\|}}{C}-NH_2}$

Practically neutral, unless additional —CO— or other electronegative group is attached

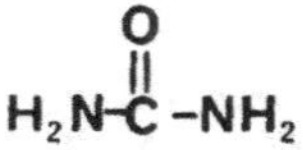

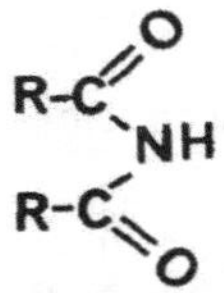

Phenobarbital

Acidic

Acidic

Imides

$\overset{\displaystyle R-\overset{\overset{\displaystyle O}{\|}}{C}}{\underset{\displaystyle R-\underset{\underset{\displaystyle O}{\|}}{C}}{\big\rangle NH}}$

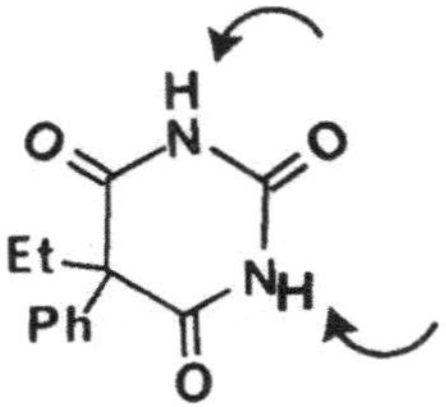

Weakly acidic

7.41

Phenytoin

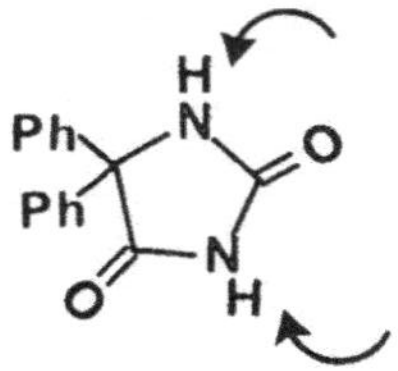

Neutral

Acidic

ca. 7.4

71

TABLE 11 (Continued).

Functional Group	Category	pKa	Comments	Examples
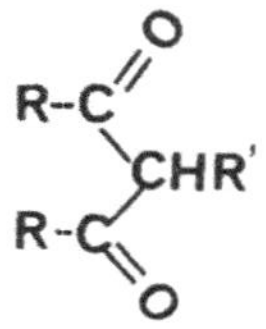 β-Diketones	Weakly acidic	4.5	Due to in- ductive effect of two C=O groups; keto-enol tautomerism may exist under high pH	Phenylbutazone 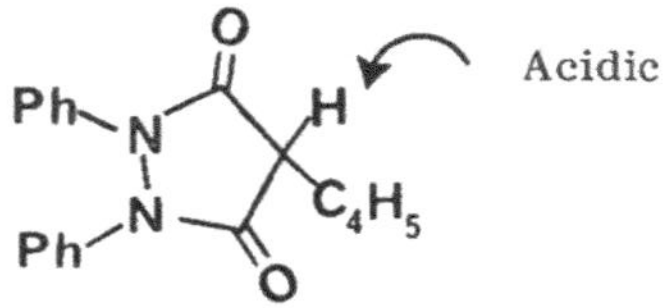 Acidic
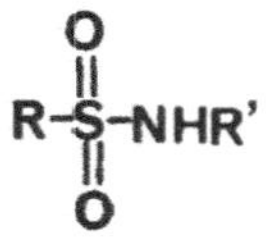 Sulfonamides	Weakly acidic to acidic	5.0	Due to the powerful in- ductive effect of —SO₂— group and that of R'	Sulfisoxazole 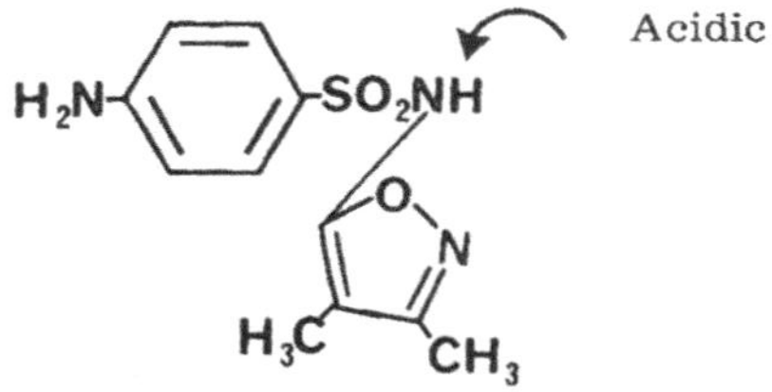 Acidic
		5.43		Tolbutamide 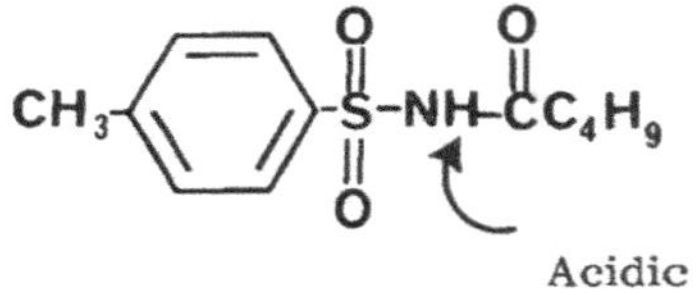 Acidic

R-SO₂NHR'

Weakly acidic to acidic

6–10

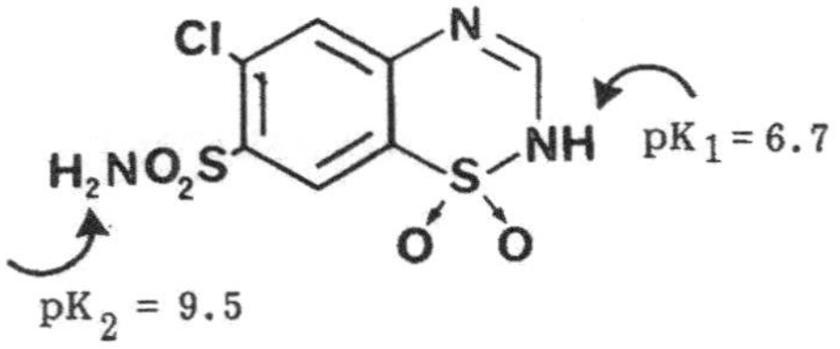

pK₁ = 6.7

pK₂ = 9.5

Chlorothiazide

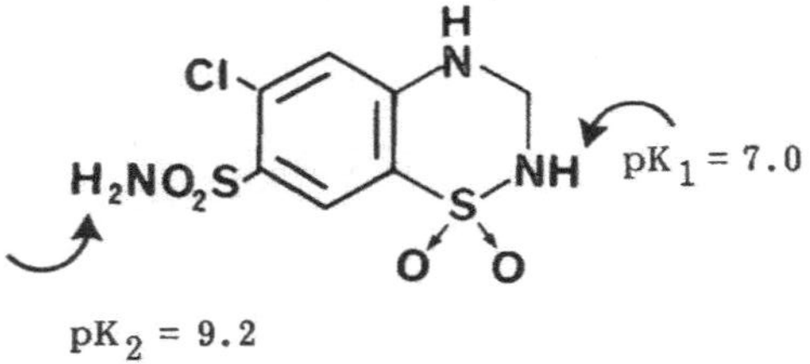

pK₁ = 7.0

pK₂ = 9.2

Hydrochlorothiazide

Exact pKₐ depends on both R and R'

ArNH₂
ArNHR } Aromatic amines
ArNR₂

Weakly basic

2.78

Benzocaine

Ranges from 3 to 6; influenced by resonance and inductive effects

ArNAr pKₐ ca. 1
          |
          H

ArN(Ar)₂ is neutral

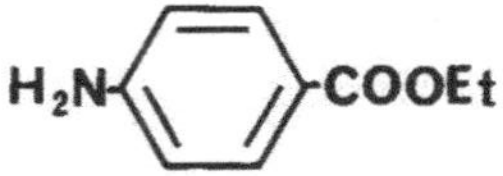

73

TABLE 11 (Continued).

Functional Group	Category	pK$_a$	Comments	Examples
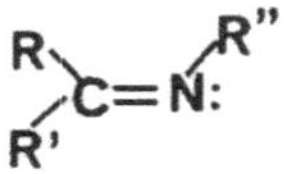 R\C=N:R" R' Imines	Weakly basic	3.4		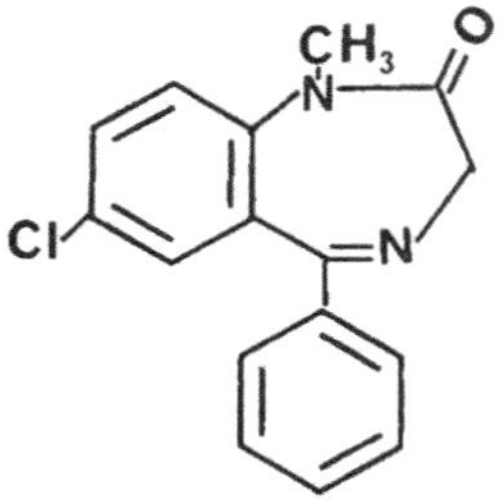 Diazepam
RCOOH Carboxylic acid	Acidic	4.2	Influenced by inductive and resonance effects	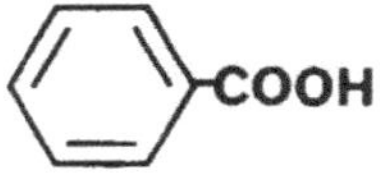 Benzoic acid
		2.73		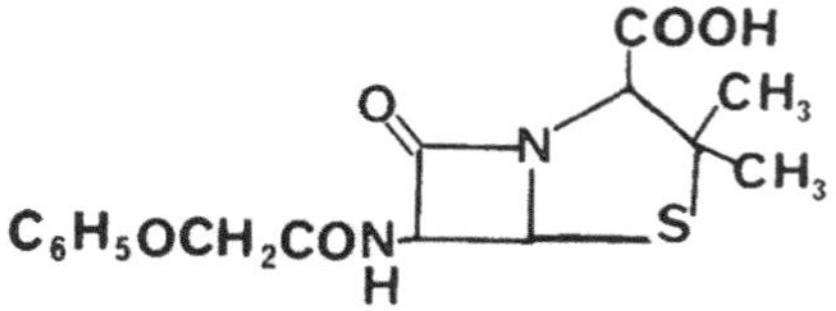 Penicillin V

Diazepam

Benzoic acid

Penicillin V

Structure	Class	Acidity	pK_a	Example
RCOOR'	Esters	Neutral		Too weak to be measured in water 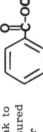 Benzyl benzoate
H–C≡		Neutral to extremely weak acid	9.14	H–C≡N Hydrocyanic acid (highly toxic)
H–C=C	Olefins	Extremely weak acid	>10	
Saturated hydrocarbons		Neutral		Petrolatum, mineral oil

TABLE 12 Strengths of Acids and Bases

Class	Dissociation constant, K	Dissociation exponent, pK (pK_a for acids, pK_b for bases)
Strong	$>10^{-2}$	<2[a]
Intermediate	$10^{-2}-10^{-7}$	$2-7$
Weak	$10^{-7}-10^{-12}$	$7-12$[b]
Feeble	$<10^{-12}$	>12

[a]For a strong base, since the pK_b is < 2, the pK_a of the conjugate acid will be >12.
[b]For a weak base, since the pK_b is $7-12$, the pK_a of the conjugate acid will be $7-2$.

c. Temperature, exothermic vs. endothermic

$$\log S = \frac{-\Delta H}{2.303R} \frac{1}{T} + \text{constant}$$

$$\log \frac{S_2}{S_1} = \frac{\Delta H}{2.303R} \left(\frac{1}{T_1} - \frac{1}{T_2} \right)$$

Compare these equations with the van't Hoff and Clausius-Clapeyron equations

d. Pressure, important for gases; Henry's relationship: $C_2 = kp$.

e. Particle size and agitation usually affect only the rate of dissolution, but in some special cases may affect the solubility (e.g., gypsum, $CaSO_4$).

3. *Rate of Solution and Fick's Law of Diffusion (Figure 8)*

Rate of solution $= \frac{DA}{\ell} (C_1 - C_2) = \frac{DA}{\ell}$ (saturated solubility)

Bulk solution

D = diffusion coefficient
A = surface area
ℓ = thickness of the stagnant layer
C_1, C_2 = concentrations

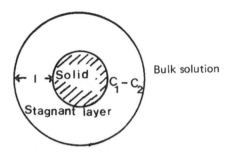

Figure 8 Graphic presentation of the stagnant layer of solvent surrounding a solid particle.

4. Official (USP, NF) Solubility Expressions

The solubility of a drug can be expressed in many different ways. For the convenience of dispensing pharmacists, the *U. S. Pharmacopeia* and the *National Formulary* define the solubility of drugs as the number of milliliters of solvent in which 1 g of solute will dissolve at 25°C (23) (Table 13).

TABLE 13 Official Solubility Expressions of Drugs

Term	Parts of solvent required for 1 part of solute
1. Very soluble	<1 part
2. Freely soluble	1–10 parts
3. Soluble	10–30
4. Sparingly soluble	30–100
5. Slightly soluble	100–1000
6. Very slightly soluble	1000–10,000
7. Practically insoluble (insoluble)	>10,000

5. Solubility Rules: Influence of Functional Group

"Like dissolves like," polar group in general increases the solubility of a compound in water (Table 14). Why resorcinol has higher water solubility than pyrocatechol and pyrogallol can be attributed to intramolecular hydrogen bonding of the vicinal (ortho-) OH groups, thus reducing hydrogen bonding with H_2O. The meta-dihydroxy groups are too far to form an intramoelcular hydrogen bond, and therefore can interact with water.

TABLE 14 Influence of Polar Groups on the Solubility of Organic Compounds

Compound	Parts of H_2O required for 1 gm of the solute	Official expression
C_6H_6 (benzene)	1430	Very slightly soluble
C_6H_5COOH	276	Slightly soluble
$C_6H_5NH_2$	28.6	Soluble
$C_6H_5CH_2OH$ (benzyl alcohol)	25	Soluble
C_6H_5OH (phenol)	15	Soluble
$1,4\text{-}C_6H_4(OH)_2$ (hydroquinone)	14	Soluble
$1,2\text{-}C_6H_4(OH)_2$ (pyrocatechol)	2.3	Fairly soluble
$1,2,3\text{-}C_6H_3(OH)_3$ (pyrogallol)	1.7	Fairly soluble
$1,3\text{-}C_6H_4(OH)_2$ (resorcinol)	0.9	Very soluble

6. *Solubilities of Weak Electrolytes: Influence of Functional Groups*

Basic compounds such as RNH_2, $RNHR'$, R_3N, and $ArNH_2$ are more soluble in acidic solutions than in neutral or basic solutions; acidic compounds such as RCOOH and Ar—OH are more soluble in basic solutions than in acidic solutions (Scheme 14). The exact degree of solubility will depend on the molecular weight and the presence of other moieties, branching, degree of symmetry, H bonding, etc.

C_6H_5OH

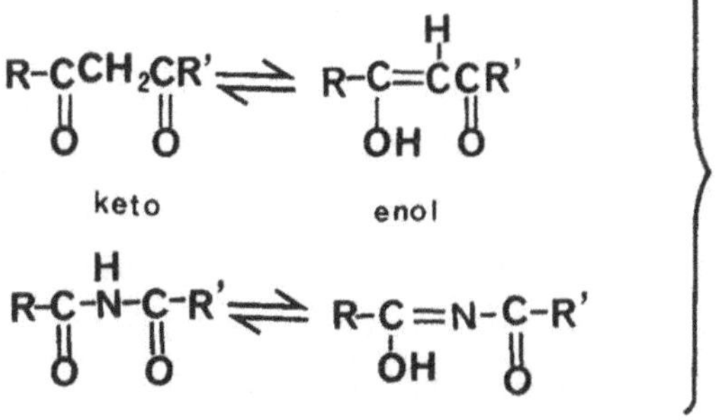

Phenolate more water soluble than phenol

$$R-\underset{\underset{O}{\|}}{C}CH_2\underset{\underset{O}{\|}}{C}R' \rightleftharpoons R-\underset{\underset{OH}{|}}{C}=\overset{H}{C}\underset{\underset{O}{\|}}{C}R'$$

keto   enol

Weak acids, more soluble in basic solution; tautomerism exists, expecially under basic conditions

$$R-\underset{\underset{O}{\|}}{C}\overset{H}{-}N-\underset{\underset{O}{\|}}{C}-R' \rightleftharpoons R-\underset{\underset{OH}{|}}{C}=N-\underset{\underset{O}{\|}}{C}-R'$$

Barbiturates

Scheme 14

General rules:

Weak electrolytes of moderate to high molecular weight ($>C_5$)   at different pH values

Ionic: behave like strong electrolytes; more soluble in water than in semipolar or nonpolar solvents

Nonionic: behave like nonelectrolytes; not soluble in water; more soluble in semipolar solvents such as EtOH

7. *Solvent–Solute Interactions*

According to their relative polarity (dielectric constants), solvents can be grouped into three categories:

1. *Polar solvents*: water (e = 80), dilute acids and dilute bases, hydroalcoholic solvent with low alcohol content
2. *Semipolar solvents*: ketones, alcohols (e = 50–20)
3. *Nonpolar solvents*: benzene, petroleum ether, carbon tetrachloride (e < 5)

Why water is a good solvent for salts, sugars, strong electrolytes, amino acids, and polyhydroxy compounds can be explained as follows:

1. High dielectric contant of H_2O reduces interionic forces from 100–200 kcal/mol to about 5 kcal/mol.
2. H_2O breaks covalent bond such as H—Cl by acid-base reactions, and stabilization of the ions by hydration.
3. H bonding accounts for the high solubility of sugars, alcohols, aldehydes, and other O- and N-containing compounds of low molecular weight (less than C_5 for each functional group).
4. Ion-dipole interaction accounts for the solubility of soap and salts in water.

Nonpolar solvents are poor solvents for ionic compounds but good for oils, fats, alkaloid free bases, and fatty acids ($>C_5$) for the following reasons:

1. Unable to reduce interionic forces due to low e, and unable to form H bonds; therefore, they are poor for ionic compounds.

2. The weak van der Waals forces (induced dipole—induced dipole) are additive (0.5 to 1 kcal/mol/CH$_2$) and will enable them to dissolve nonpolar compounds (e.g., oils, fats, undissociated bases, and acids).

Pharmaceutical examples of solubility problems and incompatibilities are shown by the effect of different pH values on the solubility of phenobarbital and sulfadiazine (Schemes 15 and 16).

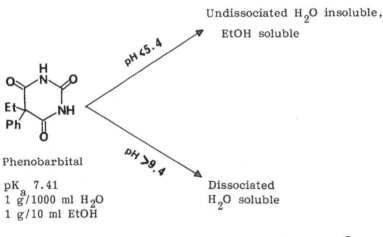

Undissociated H$_2$O insoluble, EtOH soluble

pH <5.4

pH >9.4

Dissociated H$_2$O soluble

Phenobarbital

pK$_a$ 7.41
1 g/1000 ml H$_2$O
1 g/10 ml EtOH

keto

enol

Phenobarbital sodium
1 g/ml H$_2$O
Freely soluble in water

Scheme 15

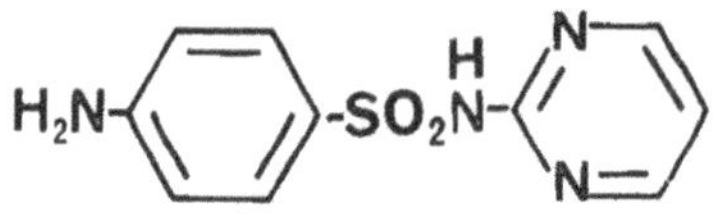

$$pK_a \ 6.48$$

Sulfadiazine

1 g/13,000 ml H_2O — may crystallize in the renal tubles and cause hematuria and anuria

+ NaHCO$_3$ — alkalizes the urinary pH and thus prevents adverse renal reactions

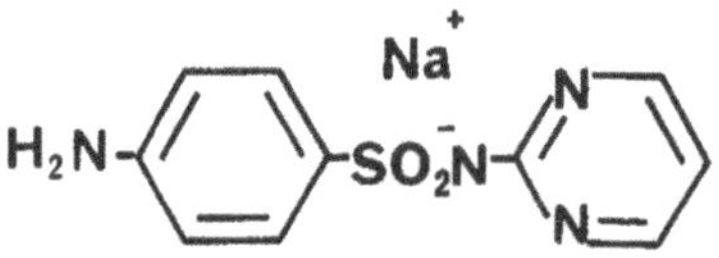

Sodium sulfadiazine

1 g/2 ml H_2O, freely soluble

Scheme 16

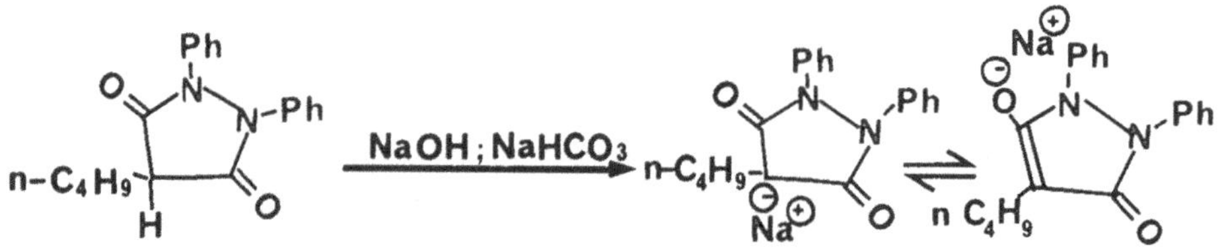

Phenylbutazone (butazolidin)   Sodium salt

Very slightly soluble in H_2O, 0.7 g/100 ml   Freely soluble in H_2O

Scheme 17

Different salts of alkaloids may have quite different solubil-ities; for example, codeine phosphate has the highest water solubility and is the form most commonly used in cough syrups (Table 15).

TABLE 15 Componison of the Solubilities of Different Salts of Codeine in Water and in Ethanol

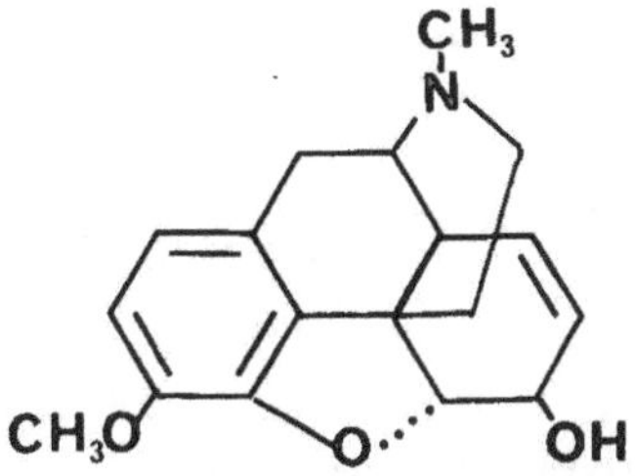

Codeine

Form	Solubility in H_2O	Solubility in EtOH
Free base	1 g/120 ml Sl. soluble	1 g/2 ml *Freely soluble*
Codeine $\cdot H_2SO_4 \cdot 5H_2O$	1 g/30 ml Soluble	1 g/1280 ml sl. soluble
Codeine $\cdot HCl \cdot 2H_2O$	1 g/20 ml Soluble	1 g/180 ml Sl. soluble
Codeine $\cdot H_3PO_4 \cdot 1\frac{1}{2}H_2O$	1 g/2.3 ml *Freely soluble*	1 g/325 ml Sl. soluble

8. Exceptions to the Rules

Some solvents have dielectric constants over 80 and yet are good solvents for amides, phenobarbital, and urea derivatives:

Formamide, $HCONH_2$, e = 109 (20°C)

N-Methylformamide, e = 190, $HCONHCH_3$

N,N-Dimethylformamide (DMF), $HCON(CH_3)_2$, e = 36.7 (25°C), μ = 3.82 D; known as a "universal solvent" in organic laboratories and industry

This is due to high permanent dipole moments, H bonding, and polarizability of these solvents.

Ephedrine, a relatively low molecular weight alkaloid from *Ephedra* with two polar functional groups, is soluble in water, alcohol, ether, chloroform, and oils

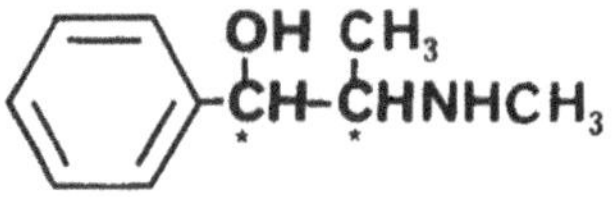

d,l form (racenic)   (19)

The HCl salt of ephedrine is freely soluble in H_2O (1 g/4 ml) and sparingly soluble (1 g/40 ml) in 95% alcohol at 20°C.

9. *Solubility Interactions*

a. *Salting out.* Addition of a salt lowers the solubility of nonpolar compounds in water (e.g., aromatic waters may turn cloudy due to oil separation upon the addition of NaCl) or a salt dissolved in water may precipitate as fine crystals upon the addition of alcohol or other semipolar solvents.

b. *Common ion effect and the solubility porduct of slightly soluble electrolytes.*

$$B_mA_n \rightleftharpoons mB^{n+} + nA^{m-} \qquad K_{sp} = [B^{n+}]^m[A^{m-}]^n$$

According to the law of mass action and Le Chatelier's principle, addition of NaCl will decrease the solubility of AgCl ($K_{sp} = 1.25 \times 10^{-10}$). Table 16 summarizes the K_{sp} of several different inorganic compounds.

TABLE 16 Solubility Products (K_{sp}) of Several Inorganic Compounds with Low Solubilities

Substance	K_{sp}	Temperature (°C)	Comment
Al(OH)$_3$	7.7×10^{-13}	25	
BaSO$_4$	1×10^{-10}	25	Nontoxic; used in roentgenographic examination of the lower bowel; soluble Ba^{2+} salts are toxic
Fe(OH)$_3$	1×10^{-36}	18	
Fe(OH)$_2$	1.6×10^{-14}	18	
AgI	1.5×10^{-16}	25	

D. **Quantitative Treatment of the Acid-Base and Solubility Equilibria**

1. *Henderson-Hasselbach Equation (Buffer Equation)*

$$pH = pK_a + \log \frac{[B]}{[A]}$$ can be used for either an acid or a base

where B = base, or the conjugate base of an acid

A = acid, or the conjugate acid of a base

For acids: HA $\rightleftharpoons$ H$^+$ + A$^-$

$$K_a = \frac{[H^+][A^-]}{[HA]}$$ taking the negative log, we have

$$pH = pK_a + \log \frac{[A^-]}{[HA]}$$ which can be expressed in any of the following forms for various applications

$$pH = pK_a + \log \frac{[salt]}{[acid]}$$ for making buffer solutions

$$pH = pK_a + \log \frac{[i]}{[u]}$$ for calculating ionized/nonionized ratio

$$pH = pK_a + \log \frac{(S - S_o)}{(S_o)}$$ where S is the total solubility and S_o the solubility of the undissociated form

For bases: $H:B^+ \rightleftharpoons H^+ + :B$ One can derive the buffer equation by using the pK_a of the conjugate acid of the base

$$K_a = \frac{[H^+][B]}{[HB^+]}$$ taking the negative log of both sides, we have

$$pH = pK_a + \log \frac{[B]}{[HB^+]}$$

$$pH = pK_a + \log \frac{[base]}{[salt]}$$

$$pH = pK_a + \log \frac{(S_o)}{(S - S_o)}$$ N.B.: Compare the subtle differences between these equations and those for acids

$$pH = pK_a + \log \frac{[u]}{[i]}$$

Calculation of the degree of ionization α and percent ionized for acids and bases is shown as follows:

$$HA \rightleftharpoons H^+ + A^- \qquad \alpha = \frac{[A^-]}{[HA] + [A^-]}$$

$$pK_a - pH = \log \frac{[HA]}{[A^-]}$$ from the buffer equation

Therefore,

$$\frac{[HA]}{[A^-]} = \text{antilog}(pK_a - pH) \qquad \text{adding 1 to both sides, we have}$$

$$\frac{[A^-] + [HA]}{[A^-]} = 1 + \text{antilog}(pK_a - pH)$$

reciprocal: $\dfrac{[A^-]}{[HA] + [A^-]} = \dfrac{1}{1 + \text{antilog}(pK_a - pH)}$

$$\alpha = \frac{1}{1 + \text{antilog}(pK_a - pH)} = \frac{1}{1 + 10^{(pK_a - pH)}}$$

% ionized = 100% × α

By similar procedures one can derive the following equations for *basic drugs*:

$$\alpha = \frac{[BH^+]}{[BH^+] + [B]}$$

$$= \frac{1}{1 + \text{antilog}(pH - pK_a)}$$

$$= \frac{1}{1 + 10^{(pH - pK_a)}} \qquad \begin{array}{l}\text{N.B.: Compare these equations}\\\text{with the equations for acids above}\end{array}$$

% ionized = 100% × α

Table 17 gives the percent ionized for both acids and bases when the (pK_a − pH) is known. It is worth noting that when (pK_a − pH) = −2, we have approximately 99% ionized for an acid or 1% ionized for a base. When pK_a − pH = 0, we have 50% ionized for both acids and bases. On the other hand, when (pK_a − pH) = +2, we have approximately 1% ionized for an acid and 99% ionized for a base. Here the principle to remember is "opposite charges attract each other" instead of "like dissolves like."

TABLE 17 Calculation of Percentage Ionized Given pK_a and pH[a]

$pK_a - pH$	For anion (RCOO⁻) (A⁻)	For cation (RNH_3^+) (HB^+)
−6.0	99.99990	0.0000999
−5.0	99.99900	0.0009999
−4.0	99.9900	0.0099990
−3.5	99.968	0.0316
−3.4	99.960	0.0398
−3.3	99.950	0.0501
−3.2	99.937	0.0630
−3.1	99.921	0.0794
−3.0	99.90	0.09991
−2.9	99.87	0.1257
−2.8	99.84	0.1582
−2.7	99.80	0.1991
−2.6	99.75	0.2505
−2.5	99.68	0.3152
−2.4	99.60	0.3966
−2.3	99.50	0.4987
−2.2	99.37	0.6270
−2.1	99.21	0.7879
−2.0	99.01	0.990
−1.9	98.76	1.243
−1.8	98.44	1.560
−1.7	98.04	1.956
−1.6	97.55	2.450
−1.5	96.93	3.07
−1.4	96.17	3.83
−1.3	95.23	4.77
−1.2	94.07	5.93

TABLE 17 (Continued).

$pK_a - pH$	For anion (RCOO⁻) (A⁻)	For cation (RNH₃⁺) (HB⁺)
−1.1	92.64	7.36
−1.0	90.91	9.09
−0.9	88.81	11.19
−0.8	86.30	13.70
−0.7	83.37	16.63
−0.6	79.93	20.07
−0.5	75.97	24.03
−0.4	71.53	28.47
−0.3	66.61	33.39
−0.2	61.32	38.68
−0.1	55.73	44.27
0	50.00	50.00
+0.1	44.27	55.73
+0.2	38.68	61.32
+0.3	33.39	66.61
+0.4	28.47	71.53
+0.5	24.03	75.97
+0.6	20.07	79.93
+0.7	16.63	83.37
+0.8	13.70	86.30
+0.9	11.19	88.81
+1.0	9.0	90.91
+1.1	7.36	92.64
+1.2	5.93	94.07
+1.3	4.77	95.23
+1.4	3.83	96.17
+1.5	3.07	96.93

TABLE 17 (Continued).

pK_a − pH	For anion (RCOO⁻) (A⁻)	For cation (RNH_3^+) (HB^+)
+1.6	2.450	97.55
+1.7	1.956	98.04
+1.8	1.560	98.44
+1.9	1.243	98.76
+2.0	0.990	99.01
+2.1	0.7879	99.21
+2.2	0.6270	99.37
+2.3	0.4987	99.50
+2.4	0.3966	99.60
+2.5	0.3152	99.68
+2.6	0.2505	99.75
+2.7	0.1991	99.80
+2.8	0.1582	99.84
+2.9	0.1257	99.87
+3.0	0.0991	99.90
+3.1	0.0794	99.921
+3.2	0.0630	99.937
+3.3	0.0501	99.950
+3.4	0.0398	99.960
+3.5	0.0316	99.968
+4.0	0.0099990	99.9900
+5.0	0.0009999	99.99900
+6.0	0.0000999	99.99990

a% ionized = α × 100%; when pK_a − pH = −2, we have approximately 99% ionized for an acid, 1% ionized for a base.

2. Effect of Branching and Molecular Symmetry on Solubility

In general, branching increases solubility, and greater symmetry increases intermolecular forces (melting point) and decreases the solubility:

n-C_4H_9OH: 1 g/12 parts of H_2O

t-Butanol $(CH_3)_3COH$ is miscible with water in any ratio

Odd carbon dicarboxylic acids in general have lower melting points and higher water solubilities than the even carbon di-carboxylic acids (Figures 9 and 10).

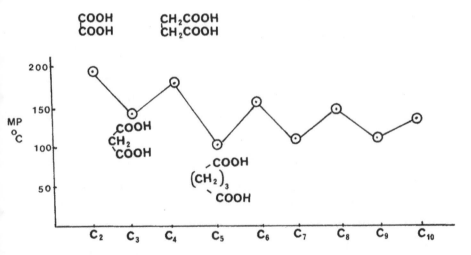

Figure 9   Plot of the melting points of dicarboxylic acids as a function of the chain length.

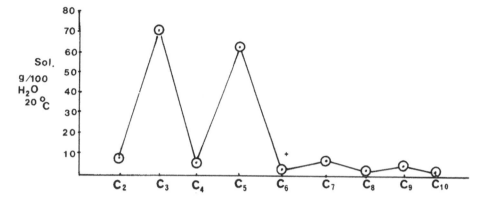

Figure 10 Plot of the solubilities of the dicarboxylic acids as a function of the chain length.

IV. MULTIPLE REGRESSION ANALYSIS AND NUMERICAL PRESENTATION OF PHYSICOCHEMICAL PARAMETERS

Over the last two decades, a number of advances have been made in quantitative structure-activity correlation studies. Hansch and co-workers (24, 25) have made the most significant contributions in this area. For a more comprehensive discussion of this topic, readers should consult review articles and monographs (26–35).

Basically, the Hansch approach is an extension of Hammett's equation (8), which has been quite successful in predicting numerous organic chemical reactions but inadequate for correlating the biological activities of highly heterogeneous biological systems.

To account for the transport and partitioning processes through biological membranes and various compartments, Hansch and Fujita used the octanol/water partition coefficint (log P for the whole molecule, π for part of a molecule) as a reference for quantitative comparison of the hydrophobicity of the drug.

Hansch's π constants:

$$\text{Log } P = \Sigma \pi$$

$$\pi = \log P_{(\text{substituted molecule})} - \log P_{(\text{parent molecule})}$$

$$\pi_{CH_3} = \log P_{CH_3 C_6 H_5} - \log P_{C_6 H_6} = 2.69 - 2.13 = 0.56$$

It is interesting to point out that Salame et al. in 1961—1962 derived the permachor method for calculation of the P factor:

$$\log P_f = 16.55 - \frac{3700}{T} - 0.22\pi$$

where π is the permachor constant. A more general equation is $\log P_f = K - R\pi$, where K is a temperature correction constant and R is a polymer (e.g., plastic) correction term. These equations are quite simlilar both in appearance and in principle to Hansch's lipophilicity constant π, although they were derived for the prediction of organic substances through plastic containers (36, 37).

Hansch et al. first derived the mathematical model, which suggested that biological activity should, in general, depend parabolically on the log partition coefficient. The classical Meyer-Overton concept and Ferguson's postulate of linear dependence can be considered as special cases, where only a limited range of log P is examined. The mathematical model used in the regression analysis can be expressed as the following:

log(biological response) = f(hydrophobicity, electronic character, steric factor)

or

$$\log \frac{1}{C} = -k_1 (\log P)^2 + k_2 \log P + k_3 \sigma + k_4 E_s + k_5$$

where C is the molar concentration required to give a standard biological response (e.g., ED_{50}, LD_{50}, etc.), P the octanol/water partition coefficient, σ the Hammett constant (a measure of the electronic effect of the substituent) and E_s the Taft steric constant. Other substituents, such as the Taft polar constant σ^*, electron density, dipole moment, pK_a, molar volume, or E_s^c, may be used. The coefficients k_1, k_2, and so on, are derived via the method of least squares, usually using a computer program

The derivation of this general equation is as follows:

$$\frac{d(BR)}{dt} = ACk_x \qquad A = f(\log P) = a \exp \left[\frac{-(\log P - \log P_0)^2}{b} \right]$$

$$= a \exp \left[\frac{-(\log P - \log P_0)^2}{b} \right] Ck_x$$

By taking the log of this equation and combining the following constants, we obtain the final equation:

$$\frac{d(BR)}{dt} = \text{const.} \qquad \log P_0 = \text{const.} \qquad \log \frac{k_x}{k_H} = \rho \sigma$$

$$\log \frac{1}{C} = -k(\log P)^2 + k' \log P + \rho \sigma + k''$$

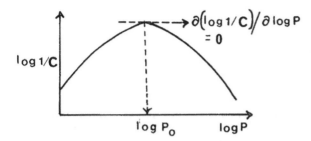

Figure 11 Parabolic dependence of log 1/C where the apex is defined as log P_0.

The ideal lipophilicity (log P_0) for maximum log 1/C is obtained by setting $\partial(\log 1/C)/\partial \log P = 0$ and solving for log P (see Figure 11). Using this approach, Hansch and various investigators have been able to obtain startling correlations of structure (expressed in numbers) with activity in a wide variety of in vitro and in vivo biological tests.   The correlation obtained would not only provide clues for further molecular modification but would also prove or disprove hypotheses proposed to explain observed facts. A few examples will serve to illustrate the usefulness of this approach.

ED_{50} for anesthesia of mice by aliphatic ethers (38):

n	r	s
26	0.97	0.10

$$\log \frac{1}{C} = -0.22(\log P)^2 + 1.04 \log P + 2.16$$

ideal log P_0 value for maximum activity = 2.35

Binding of penicillins by human serum (39):

	n	r	s

$$\log \frac{bound}{free} = 0.49\pi(\text{side chain}) - 0.63 \qquad 79 \qquad 0.92 \qquad 0.13$$

Binding of organic compounds by bovine serum albumin (pH 7.4, 37°C) (40):

$$\log \frac{1}{C} = 0.67 \log P + 2.60 \qquad\qquad 25 \qquad 0.95 \qquad 0.24$$

where n is the number of compounds examined, r the correlation coefficient, an r of 1.0 respresents a perfect correlation, and s is the standard deviation. These correlations reflect the importance of the lipophilic character in determining CNS activity and degree of protein binding.

Antagonism to epinephrine by 2-bromoalkylamines in the cat (41):

$$\log \frac{1}{C} = 1.11E_s^c + 3.57\sigma^* - 4.43n_H + 11.91 \qquad 10 \qquad 0.99$$

Antagonism to nerepinephrine (41):

$$\log \frac{1}{C} = 1.12E_s^c + 3.84\sigma^* - 4.49n_H + 11.86 \qquad 10 \qquad 0.95$$

In these equations E_s^c is Hancock's pure steric constant, σ^* is the polar constant for an aliphatic system, and n_H is the number of hydrogens on the protonated amine. Since these two equations are virtually identical, the mechanism of antagonism of both epinephrine and nerepinephrine must be the same. For a more detailed discussion of the physical meaning of these equations, the original article should be consulted (41).

During the last few years, we have reexamined the physico-chemical parameters commonly used in quantitative structure-activity relationship (QSAR) studies, as well as the various mathematical models (42—47). Our findings can be summarized as follows:

1. With the exceptions of heavy metals and polyhalogenated groups, log MR, log MW, and log MV are all well cor-related. Therefore, they can all be used as a crude measure of "bulk" in SAR (42).

	n	r	s
log MR = 0.981 log MV − 0.290	213	0.943	0.086
log MR = 0.884 log MW − 0.358	213	0.917	0.104

2. The use of log MW or the other interdependent terms is in keeping with the Einstein-Sutherland equation (43):

$$D = \frac{RT}{6\pi N \eta} \sqrt[3]{\frac{4\pi N}{3\nu MW}}$$

η = viscosity

ν = partial specific volume

π = 3.1416

When T and η are held constant, log D = const. $-1/3$ log MW $- 1/3$ log ν.

3. Different buffer species may affect the true partition coefficients (log P_{corr}), especially when the drug exists in ionized form in a substantial quantity. There-fore, there is a danger in using log P_{corr} values ob-tained from different buffer systems without correction. We have found a range of a from 0.42 to 1.16 and of b

from -1.12 to $+1.14$ for the following general equation to be used for different buffers:

$$\log P_{oct/w} = a \log P_{oct/buffer} + b$$

4. Among the different mathematical models, such as the parabolic model (48), the curvilinear model (49), the bilinear model (50), the Hyde model (51), the Higuchi-Davis equilibrium model (52), the asymptotic model of Ho (53), and the Lien model (44), the last one has been shown to be mathematically simpler to use, and the parameters involved are easier to interpret in terms of physical meaning. For pharmacokinetically controlled processes,

$$\log A = -k_1(\log P)^2 + k_2(\log P) + m \log \frac{U}{D} + n(\log MW)$$
$$+ q\mu + \chi + k_3$$

where A is the activity (1/C) or absorption (% absorption or k), $\log U/D = pK_a - pH$ for acids, μ is the dipole moment, and χ represents branching or other steric factors.

The different effects of the log MW term on the overall shape of the log A—log P plot are shown in Figure 12. It is not surprising to see that when log MW term is not included, various deviations from the typical parabolic dependence on log P may occur (54).

For drug-receptor interaction, it has been examined on theoretical grounds that dipole moment may in some cases be a

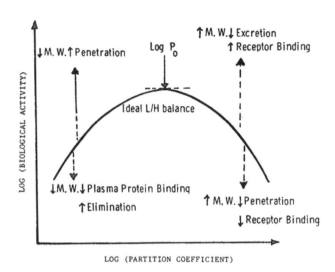

Figure 12 Possible effects of molecular weight on biological activity. MW, molecular weight; L/H, lipophilic/hydrophilic; P_O, ideal partition coefficient. The long arrows indicate the direction in which the parabola will be shifted as a result of molecular weight changes. (Adapted from Ref. 54, with permission from *J. Parent. Sci. Technol.*)

better parameter than σ to use for linear-free energy-related interactions since σ is derived from *intramolecular* electronic effect rather *intermolecular* effects (45). We have compiled the dipole moments of 306 aromatic substituents (45) and 214 aliphatic substituents (55) for future use (see Appendixes I and and II). For groups not listed, either calculated values (based on vector summation) or (preferably) experimentally determined values should be used. Sixteen examples using dipole moment, polarizability, or ionization potential on QSAR have been reviewed. It is hoped that more applications of this parameter in both aliphatic and aromatic systems can be found in the future.

A. Parameter Selection, Transformation, and Scaling

Since statistics in general does not provide direct proof of any causal relationship, one can only hope to make an educated guess with respect to the physical meaning of any correlation obtained from the analysis of biological data and molecular structure or physicochemical properties. Because the exact nature of drug-receptor interaction is not known in many cases, it becomes a subjective matter to scale or transform a certain parameter. For example, many investigators have chosen to scale the molar refraction (electronic polarization) by 0.1, while our group prefers to use the logarithmic transformation. Our selection is based on the following:

1. The dipole momoment of the drug molecule is linearly related to the interaction energy if the drug-receptor interaction is *ion-dipole* in nature:

$$E = \frac{-N_a e\mu \cos\theta}{D(r^2 - d^2)}$$

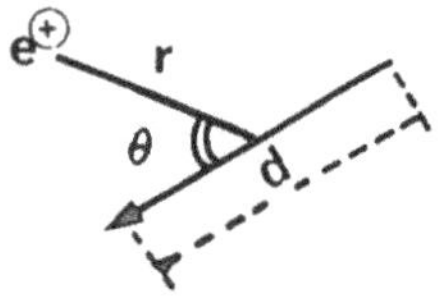

where N_a is Avogadro's number, e the magnitude of the charge on the receptor, θ the angle between the line connecting the charge and the center of the dipole and the line of the dipole moment, D the dielectric constant, r the distance between the charge and the center of the dipole, and d the length of the dipole (45).

2. If the interaction is of *dipole-dipole* type for the most favorable alignment one has

$$E = \frac{2\mu_a\mu_b}{Dr^3}$$

Since the relationship between the dipole moment μ and the molar refraction (induced or electronic polarization) of a drug molecule is given by (56).

$$\mu = 0.0128 \sqrt{(P - P_i)T}$$

where P is the total polarization,

$$P_i = \frac{4}{3}\pi N_a\alpha = \frac{n^2 - 1}{n^2 + 2}\frac{MW}{d} = MR$$

and MW/d = MV, one should use either $\sqrt{MR}$ or log MR in the correlation, in order to maintain a *linear* free energy relationship.

If, on the other hand, the drug-receptor interaction involves *all orientation* (not very likely for any specific type of interaction), the average interaction energy becomes.

$$E = \frac{-2\mu_a^2\mu_b^2}{3kTDr^6}$$

where k is Boltzmann's constant.

Alternatively, if the interaction is of the *dipole-induced dipole* (Debye forces) type, one has

$$E = \frac{-\mu_a^2\alpha_a + \mu_b^2\alpha_b}{D^2r^6}$$

where α_a and α_b are the polarizabilities of the drug and the receptor groups, respectively. Only under these two circumstances can one justify the use MR or 0.1 MR, since μ^2 is proportional to MR.

Our group has examined the interrelationships among MR, MW, and MV (molar volume) (42). It is interesting to point out that there is a high degree of collinearity among these parameters, especially if one excludes heavy groups such as polyhalogenated groups and metals (Table 18). It is noteworthy that the log scale gives better correlations than 0.1MR.

We have also examined the apparent correlations between molecular weight and therapeutic dose of drugs, using both log 1/C (mol/kg) and log D (mg/kg) (Table 19).

One can argue that dose (mg/kg) = (number of molecules × molecular weight)/kg; therefore, by using log D (mg/kg), one may introduce the log MW term into the equation. Furthermore, from the biochemical point of view, it is more meaningful to use (number of molecules)/kg instead of weight (mg)/kg, even though that is the commonly expressed dosage (57).

V. QSAR OF CONVULSANTS, ANTICONVULSANTS, AND DEPRESSANTS

Although knowledge of the receptors of CNS-acting drugs is not quite as complete as that of the nicotinic cholinergic receptor, considerable progress has been made in the last few years regarding the complexity of the receptor system for many CNS-acting drugs. Figure 13 shows a schematic diagram of various CNS-acting drugs interacting at the benzodiazepine-γ-aminobutyric acid (GABA) receptor- chloride ionophore system (58).

TABLE 18 Equations Correlating log MR with log MV, log MW, and 0.1MW

n	r	r^2	s	Equation	Equation number
266	0.913	0.833	0.109	log MR = −0.240 + 0.953 log MV	1
266	0.778	0.605	0.169	log MR = −0.211 + 0.776 log MW	2
266	0.704	0.496	0.923	0.1MR = 0.590 + 0.174 (0.1MW)	3
Total − (polyfluorinated + polybrominated + metals)					
213	0.943	0.889	0.086	log MR = −0.290 + 0.981 log MV	4
213	0.917	0.840	0.104	log MR = −0.358 + 0.884 log MW	5
213	0.895	0.801	0.592	0.1MR = 0.204 + 0.244 (0.1MW)	6

TABLE 19 Apparent Correlations Between log 1/C and log MW of Various Drugs

	n	r	r^2	s	Significance level (%)
Water-soluble vitamins					
log 1/C (mol/kg) = 3.830 log MW − 2.630	8	0.922	0.851	0.549	99.5
log D (mg/kg) = −2.826 log MW + 5.628	8	0.870	0.756	0.550	99.5
CNS-acting drugs					
log 1/C (mol/kg) = 5.051 log MW − 6.756	15	0.905	0.820	0.719	99.95
log D (mg/kg) = −4.043 log MW + 9.760	15	0.865	0.749	0.713	99.95
Anticancer drugs					
log 1/C (mol/kg) = 3.411 log MW − 2.559	15	0.786	0.618	0.837	99.9
log D (mg/kg) = −2.401 log MW + 5.536	15	0.669	0.447	0.838	99.5

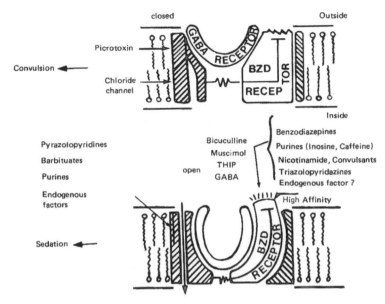

THIP = 6,7-Tetrahydroisoxazolo-5,4-c-pyridin-3-ol

Figure 13 Model of the benzodiazepine-GABA receptor-chloride inophore system. (Adapted from Ref. 58, with permission from Academic Press.)

According to this diagram, while barbiturates and picrotoxin act at the chloride inophore, the benzodiazepine receptor is more closely associated with the inhibitory GABA receptor. In the following sections specific examples of quantitative structure-activity correlations involving different pharmacological receptors will be presented.

Table 20 summarizes the equations correlating the CNS activities with the physicochemical constants of various compounds. It is apparent from these correlations that among the congeneric series of compounds examined, the relative lipophilic-

TABLE 20 Equations Correlating CNS Activity with Physicochemical Constants

Animal	Test	Compounds	Eq. no.	Equation	n	r	s	log P_o (±95% CL)[a]	Ref.
Mice	Antielectroshock	Cyclohexanones, etc.	1	$\log 1/C = -0.138\,(\log P)^2 + 0.736 \log P + 2.323$	8	0.43	0.47	2.66 (± ∞)	46
			2	$\log 1/C = -0.212\mu + 3.588$	8	0.51	0.41	—	
			3	$\log 1/C = -0.267\,(\log P)^2 + 1.423 \log P -0.368\mu + 2.619$	9	0.89	0.26	2.66 (1.64–3.17)	46
Mice	Muscle relaxant	Cyclohexanones, etc.	4	$\log 1/C = -0.237\,(\log P)^2 + 1.181 \log P + 1.591$	8	0.68	0.36	2.49 (± ∞)	46
			5	$\log 1/C = -0.313\,(\log P)^2 + 1.601 \log P -0.276\mu + 1.895)$	8	0.92	0.21	2.56 (2.08–2.85)	46
Mice	Acute lethal toxicity	Cyclohexanones, etc.	6	$\log 1/C = -0.138\,(\log P)^2 + 0.729 \log P + 1.649$	8	0.42	0.38	2.65 (± ∞)	46
			7	$\log 1/C = -0.217\,(\log P)^2 + 1.166 \log P + 0.286\mu + 1.964$	8	0.85	0.25	2.69 (± ∞)	46
Mice	Antielectroshock	Miscellaneous	8	$\log 1/C = -0.222\,(\log P)^2 + 1.153 \log P - 0.368\mu + 2.994$	18	0.92	0.24	2.59 (2.39–2.84)	46
Mice	Maximal electroshock (MES)	Miscellaneous	9	$\log 1/C = 7.776 \log MW - 14.438$	13	0.94	0.24	—	47
Mice	Maximal electroshock	1,4-Benzodiazepinones	10	$\log 1/C = -0.258\,\pi^2 + 0.361\pi + 0.954\sigma + 3.660$ $\pi_o = 0.70\ (0.08{-}5.35)$	11	0.82	0.42	—	47
Mice	Minimal electroshock	1,4-Benzodiazepinones	11	$\log 1/C = -0.220\,\pi^2 + 0.081\pi + 0.984\sigma + 3.260$ $\pi_o = 0.18\ (-2.82{-}2.37)$	10	0.83	0.33	—	47

Species	Test	Class	No.	Equation	n	r	s		Ref.
Mice	Antielectroshock	Hydantoins	12	$\log 1/C = 0.490 \log P + 0.525 E_s(R_1) + 2.286$	11	0.92	0.26	—	46
Rats	Antielectroshock	Hydantoins	13	$\log 1/C = 0.298 \log P + 0.373 E_s(R_1) + 2.727$	11	0.91	0.19	—	46
	Antielectroshock	Hydantoins	14	$\log 1/C = -0.330 (\log P)^2 + 1.124 \log P + 0.540 E_s(R_1) + 2.330$	11	0.96	0.14	1.70 (1.45—3.88)	46
Mice	Antipentylene-tetrazole-induced seizures	Barbiturates, hydantoins, imides	15	$\log 1/C = 1.023 \log P - 1.224$	5	0.98	0.21	—	46
			16	$\log 1/C = 0.998 \log P - 0.482\mu - 0.486$	5	0.996	0.116	—	46
Mice	Antipentylene-tetrazole-induced seizures	Barbiturates, hydantoins, imides	17	$\log 1/C = -0.142 (\log P)^2 + 0.695 \log P - 0.111$	13	0.93	0.22	2.44 (1.68—9.57)	46
			18	$\log 1/C = -0.145 (\log P)^2 + 0.716 \log P - 0.093$	10	0.94	0.24	2.47 ($\pm \infty$)	46
			19	$\log 1/C = -0.123 (\log P)^2 + 0.588 \log P - 5.97\mu + 0.825$	10	0.99	0.12	2.39 (1.72—5.39)	46
Mice	Antipentylene-tetrazole-induced seizures (MET)	Miscellaneous	20	$\log 1/C = -0.301 (\log P)^2 + 0.852 \log P - 0.629\mu + 4.139$	12	0.92	0.23	1.42 (1.06—9.61)	47
Mice	Antipentylene-tetrazole-induced seizures	1,4-Benzodiazepinones	21	$\log 1/C = 0.307\pi^2 + 0.144\pi + 1.291\sigma + 4.558$ $\pi_0 = 0.23 \ (-2.86-3.75)$	10	0.87	0.47		47
Mice	Rotorod ataxia	Miscellaneous	22	$\log 1/C = 15.939 \log MW - 0.972 \log P + 0.549\mu - 33.187$	16	0.93	0.39		47

TABLE 20 (Continued).

Animal	Test	Compounds	Eq. no.	Equation	n	r	s	log P_0 (±95% CL)[a]	Ref.
Mice	Acute lethal toxicity	Barbiturates, hydantoins, imides	23	log 1/C = −0.207 (log P)^2 + 0.802 log P − 0.523	15	0.84	0.37		46
			24	log 1/C = −0.210 (log P)^2 + 0.856 log P − 0.510	14	0.94	0.23		46
			25	log 1/C = −0.236 (log P)^2 + 0.877 log P − 0.381	10	0.98	0.16		46
			26	log 1.C = −0.226 (log P)^2 + 0.800 log P − 0.361μ + 0.175	10	0.99	0.11		46
Mice	Acute lethal toxicity	Lactams, thio-lactams, ureas, thioureas	27	log 1/C = −0.373 (log P)^2 + 1.010 log P + 0.227μ + 1.405	22	0.83	0.31		46
			28	log 1/C = −0.364 (log P)^2 + 1.005 log P + 0.247μ + 1.298	20	0.89	0.24		46
Mice	Acute lethal toxicity	N-Substituted lactams, o-phenyleneureas	29	log 1/C = 0.312 log P + 2.468	5	0.98	0.06		46

[a]CL, confidence limit.

ity (log P) affects the relative CNS activities. Furthermore,
the ideal lipophilicity (log P_0) for maximum activity centers
around 2.0 for the undissociated forms. The exact log P_0
value depends not only on the series of drugs but also on the
testing system employed (Table 21). Another interesting finding
is that for anticonvulsant and CNS depressant activities the
dependence on the electronic parameter μ (dipole moment) is
usually negative (equations 2, 3, 5, 7, 8, 16, 19, 20, 26),
while the convulsant activity or toxicity dependends positively
on μ (equations 21, 27, 28). This seems to suggest that different types of intermolecular forces are involved. The con-
tribution from the steric parameter is significant only in limited
cases (equations 12, 13).

VI. QSAR OF NARCOTIC ANALGETICS

Analgetics are drugs that alleviate pain without significantly
impairing consciousness. Using the bradykinin-evoked visceral
pain in the cross-perfused spleen preparation in dogs, Lim et al.
of Miles Laboraotry (59) have proposed that analgetics be
classified in three groups: (a) peripherally acting, nonnarcotic
(e.g., salicylates); (b) centrally acting, nonnarcotic (e.g.,
d-propoxyphene); and (c) centrally acting, narcotic (e.g.,
morphine, meperidine, etc.). This grouping is by no means
absolute. For example, d-propoxyphen has some peripheral
effect as well as central effect, and its dependence liability,
although low compared to morphine or meperidine, can still be
a problem on prolonged use.

The major sites of action of centrally acting narcotic analgetics are the cerebrum and medulla (60). Therefore, their

TABLE 21 Optimum Lipophilic Character (log P_O) for Maximum CNS Activity

Animal	Compounds	No. of data points	CNS Activity	log P_O ($\pm95\%$ CL)[a]	Ref.
Mice	Cyclohexanones, etc.	8	Antielectroshock	2.66 (1.64–3.17)	46
Mice	Barbiturates, hydantoins, imides, cyclohexanones, etc.	18	Antielectroshock	2.59 (2.39–2.84)	46
Mice	Cyclohexanones, etc.	8	Muscle relaxant	2.56 (2.08–2.85)	46
Mice	Barbiturates	10	Hypnotic	2.40	46
Mice	Barbiturates, hydantoins, imides	10	Antipentylenetetrazole-induced seizures	2.39 (1.72–5.39)	46
Mice	Miscellaneous anticonvulsants	12	Antipentylenetetrazole-induced seizures (MET)	1.42 (1.06–9.61)	47

Species	Class	Ref.	Activity	Value (CL^a)	Ref.
Mice	2-Imidazolidinones	15	Motor activity reduction (MDD_{50})	4.17 (3.75–5.11)	47
Mice	Barbiturates, hydantoins, imides	14	Acute lethal toxicity (CNS depression)	2.04 (1.59–3.56)	46
Mice	Barbiturates, hydantoins, imides	10	Acute lethal toxicity (CNS depression)	1.77 (1.49–2.26)	46
Rats	Hydantoins	11	Antielectroshock	1.70 (1.45–3.88)	46
Mice	Ureas, thioureas, lactams, thiolactams	20	Acute lethal toxicity (CNS toxicity)	1.38 (1.17–1.71)	46
Mice	γ-Butyrolactones	5	CNS depression	<0.71	46
Mice	2-Sulfamoylbenzoates	9,12	Antistrychnine Antielectroshock	<0.13	46

aCL, confidence limits.

ability to pass the blood-brain barrier is extremely important. It is now well recognized that this ability is intimately related to the lipophilicity of the drug as measured by lipoid/water partition coefficient (46, 47).

Although the precise mechanism of drug-receptor interaction for analgetics has not been elucidated, several common structural features have been recognized: (a) a quaternary carbon atom, (b) a phenyl or isostere ring attached to this carbon atom, (c) a tertiary amino group two carbon atoms removed, and (d) a phenolic hydroxyl group in meta position relative to the point of attachment to the quaternary carbon atom if the tertiary nitrogen is part of six-membered ring (60). It becomes apparent that in addition to the lipophilicity of the drug molecule, both electronic and steric parameters are important for selective drug-receptor interaction. The main purpose of this section is to examine how the physicochemical properties of the compounds—partition coefficient, molecular weight, molar refraction, and steric constants—may affect the analgetic potency of their binding with the receptor.

The first set of data was from the work of Buckett (61) on esters of 14-hydroxycodeinone. The relative activity (morphine = 1) was converted to the logarithmic scale in order to be in line with linear-free-energy-related parameters such as log P (Table 22).

Figure 14 shows the dependence of log RA on log P. Inclusion of log MW_R improved the correlation, as indicated by the higher correlation coefficient (r) and the lower standard deviation (s) (62). The log MW_R term is statistically justifiable at 90% level, as indicated by an F test ($F_{1,9}$ = 4.3; $F_{1,9}$ .90 = 3.4).

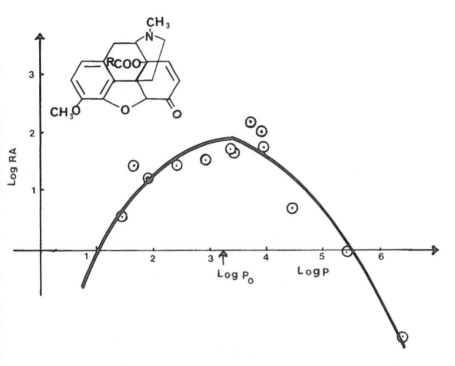

Figure 14 Parabolic dependence of the analgesic activity of esters of 14-hydroxycodeinone on log P without the log MW term.

The addition of the log MW_R term not only resulted in a better fit of the curve (see Figure 15), but also in a shift of the ideal lipophilicity (log P_O) from 3.23 to 1.45 for the unprotonated forms of these drugs.

For the analgesic activities of N-alkylnormeperidine homologs (Table 23), equations 1 to 3 were obtained (Table 24). The correlation coefficient of the parabolic equation (equation 2) is only 0.824. This is due primarily to the apparent "biphasic" curve of the log 1/C vs. log P plot, as indicated by Figure 16.

TABLE 22 Physicochemical Properties and Analgesic Activities of Esters of 14-Hydroxycodeinone in Mice

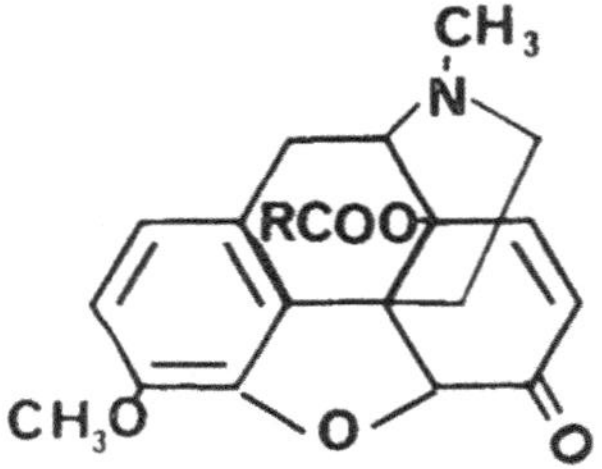

| RCO | log(relative activity) | | log P[c] | log MW(RCO) | log MR(RCO) |
	Obsd.[a]	Calcd.[b]			
CH₃CO—	0.60	0.46	1.40	1.18	0.75
C₂H₅CO—	1.27	1.24	1.90	1.46	1.01
n-C₃H₇CO—	1.46	1.60	2.40	1.63	1.17
n-C₄H₉CO—	1.59	1.74	2.90	1.76	1.29

$n\text{-}C_5H_{11}CO-$	1.67	1.67	3.40	1.85	1.38
$n\text{-}C_6H_{13}CO-$	1.78	1.45	3.90	1.93	1.46
$n\text{-}C_7H_{15}CO-$	0.71	1.13	4.40	2.00	1.53
$n\text{-}C_9H_{19}CO-$	0.05	0.03	5.40	2.10	1.63
$n\text{-}C_{11}H_{23}CO-$	-1.47	-1.51	6.40	2.19	1.72
$ph\text{-}CH_2CO-$	1.72	2.01	3.36	1.96	1.48
$ph\text{-}CH_2CH_2CO-$	2.06	1.76	3.86	2.02	1.54
$ph\text{-}CH\!=\!CHCO-$	2.25	1.91	3.66	2.01	1.53
$CH_3CH\!=\!CHCO-$	1.49	1.61	2.20	1.61	1.16

[a]Morphine = 0. Tail clip method. From Buckett (61).
[b]Calculated from equation 5.
[c]Octanol/water system, calculated from log $P_{morphine}$ = 0.76, log $P_{codeine}$ = 0.76 + 0.65 = 1.41, where 0.65 = log $P_{C_6H_5OMe}$ − log $P_{C_6H_5OH}$, π_{CH_3COO} = −0.01, π_{CH_2} = 0.50.

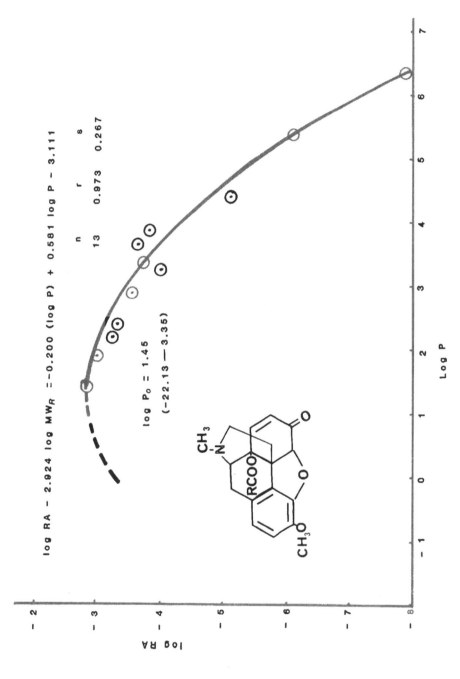

Figure 15. Parabolic dependence of the analgesic activity on log P with

TABLE 23 Physicochemical Properties and Analgesic Activities of *N*-Alkylnormeperidine Homologs

$$\text{EtOOC} \diagdown \underset{\text{Ph}}{\bigcirc} \diagup \text{N-R}$$

	Analgesia [log 1/C (mol/kg)]		Opiate receptor					
			Inhib. (log 1/C)					
			(−NaCl)		(+NaCl)			
R	Obsd.[a]	Calcd.[b]	Obsd.[a]	Calcd.[c]	Obsd.[a]	Calcd.[d]	log P_{app}	$E_s^c(R)$
CH$_3$— (meperidine)	4.48	4.49	6.30	6.18	4.40	4.51	1.28[e]	0.00
C$_2$H$_5$—	4.39	4.37	5.30	5.59	4.30	4.08	1.56	−0.38
n-C$_3$H$_7$—	4.34	4.34	5.40	5.26	4.00	3.88	1.85[e]	−0.67
n-C$_4$H$_9$—	4.70	4.73	6.05	5.89	4.52	4.62	2.13[e]	−0.70
n-C$_5$H$_{11}$—	5.00	5.03	6.40	6.44	4.82	5.23	2.41	−0.71
n-C$_6$H$_{13}$—	5.30	5.26	6.70	6.98	5.70	5.76	2.78[e]	(−0.71)[f]
n-C$_7$H$_{15}$—	5.30	5.27	7.26	7.18	6.19	5.91	3.06	(−0.71)[f]
n-C$_8$H$_{17}$—	5.00	5.08	7.52	7.20	6.10	5.77	3.43	(−0.71)[f]
n-C$_9$H$_{19}$—	4.82	4.79	6.82	7.01	5.10	5.41	3.71	(−0.71)[f]

[a]From Ref. 63.
[b]Calculated from equation 3 (Table 24).
[c]Calculated from equation 7 (Table 24).
[d]Calculated from equation 11 (Table 24).
[e]Experimentally determined in octanol-phosphate buffer, pH 7.4, from Ref. 66. The others were calculated values using π_{CH_2} = 0.28, $\pi_{CH_2CH_2}$ = 0.65.
[f]Estimated value using the E_s^c of n-C$_5$H$_{11}$.

TABLE 24 Equations Correlating Analgesic Activities with Physicochemical Constants of N-Alkyl-normeperidine Homologs

Eq. no.	Equation	n	r	s
	Analgesia (in mice, hot-plate technique)			
1	$\log 1/C = 0.306 \log P_{app} + 4.060$	9	0.703	0.278
2	$\log 1/C = -0.275(\log P_{app})^2 + 1.681 \log P_{app} - 2.515$ $\log P_O = 3.05(\pm \infty)$	9	0.824	0.239
3	$\log 1/C = -0.854(\log P_{app})^2 + 5.041 \log P_{app} + 2.246 E_s^c - 0.563$ $\log P_O = 2.95 \ (2.90-3.01)$	9	0.994	0.051
	Inhibition of [^{3}H]-naloxone binding (without NaCl)			
4	$\log 1/C = 0.718 \log P_{app} + 4.646$	9	0.796	0.491

		n		
5	$\log 1/C = 1.015 \log P_{app} + 1.436 E_s^c + 4.757$	9	0.860	0.446
6	$\log 1/C = 0.077(\log P_{app})^2 + 0.333 \log P_{app} + 5.078$	9	0.798	0.528
7	$\log 1.C = -1.075(\log P_{app})^2 + 7.015 \log P_{app} + 4.471 E_s^c - 1.043$ $\log P_O = 3.26 \ (3.01-4.85)$	9	0.959	0.271

(with NaCl)

		n		
8	$\log 1/C = 0.751 \log P_{app} + 3.161$	9	0.780	0.541
9	$\log 1/C = -0.207(\log P_{app})^2 + 1.787 \log P_{app} + 1.997$	9	0.794	0.568
10	$\log 1.C = 0.891 \log P_{app} + 0.675 E_s^c + 3.213$	9	0.794	0.509
11	$\log 1/C = -1.398(\log P_{app})^2 + 8.698 \log P_{app} + 4.622 E_s^c - 4.333$ $\log P_O = 3.11(2.91-3.97)$	9	0.947	0.328

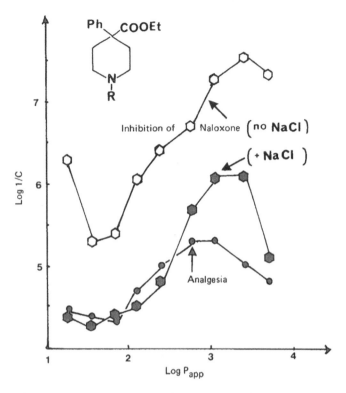

Figure 16 Dependence of the analgesic activity and receptor binding affinity on the apparent partition coefficient.

Meperidine, the lowest member in the series, appears to have activity higher than expected from the regression. However, when the method of least squares was applied, meperidine did not appear as an outlier. The deviation was only slightly higher than the standard deviation of the regression (0.264 vs. 0.239). Pert et al. (1976) have attributed the relatively high analgetic activity of meperidine to its ability to penetrate the brain (600-fold higher brain level relative to morphine) (63).

Figure 16 shows that meperidine also has a relatively high affinity in the inhibition of naloxone binding, especially in the absence of NaCl. Addition of Hancock's corrected steric constant E_s^C (1961) to the parabolic equation improved the correlation significantly, as indicated by an F test ($F_{1,5}$ = 124.96; $F_{1,5_{.9995}}$ = 63.6). Positive dependence on E_s^C suggests that the bulk tolerance on the N substituent is very small.

Figure 17 shows clearly that the E_s^C constant changes significantly from C_1 to C_3 but remains fairly constant from C_3 and above. This suggests that perhaps in addition to greater penetrability to the brain, meperidine, having the least hindered $N-CH_3$, will bind more tightly to the receptor site (presumably

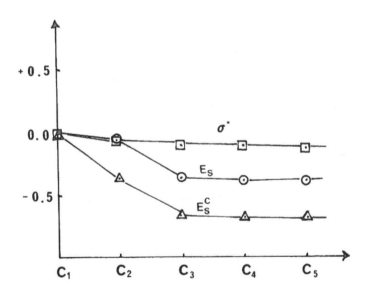

Figure 17 Dependence of the electronic and steric constants on the chain length, where $E_s^C = E_s + 0.306$ (n $-$ 3).

a negative site). This interpretation is in agreement with the finding that the receptor affinity of meperidine in the absence of sodium is more drastically enhanced than in the presence of sodium ion. The E_s^c terms in equations 7 and 11 are statistically significant at 99 and 97.5 percentile levels, respectively ($F_{1,5}$ = 17.76 for equation 9 and 12.96 for equation 11). Similar "biphasic" dependence of the analgetic activity on N-alkylnorketo-bemidones on the length of the N—R side chain has been observed by several investigators [Oh-ishi and May (64); Wilson et al. (65)]. Again this can be attributed to the difference in E_s^c constants.

The log P_o values for maximum analgesia obtained from equations 2 and 3 are around 3. This also agrees with the peak for inhibition of naloxone binding in the presence of NaCl as seen from Figure 16, as well as from equations 9 and 11.

From the analysis above, it may prove to be worthwhile to study the modification of meperidine and ketobemidone on the benzene ring or the ester or keto side chain to reach a log P of 3 (octanol/buffer pH 7.4) but to keep the N—CH$_3$ group unchanged to minimize the steric effect.

Casy (67) has analyzed the stereochemical structure-activity relationships of many C-methyl derivatives of the reversed ester of meperidine. From the results obtained it has been suggested that both the reversed ester and the meperidine groups have similar modes of interaction with the opiate receptor.

From the analgesic activities of a series of fentanyl derivatives Yang et al. (68) have derived the following correlations, using hydrophobic fragment constants. While the π of H is defined as 0.0, the f of H is 0.23. They were derived in different ways and have slightly different values for different groups (e.g., $\pi_{CH_3} = 0.50$, $f_{CH_3} = 0.77$). For a more detailed explanation the reader should consult Ref. 35 and the book by Rekker (69). From the dependence on f of the substituent L and that of R, a reasonable structural requirement of the analgesic receptor can be proposed accordingly for this type of compound (20).

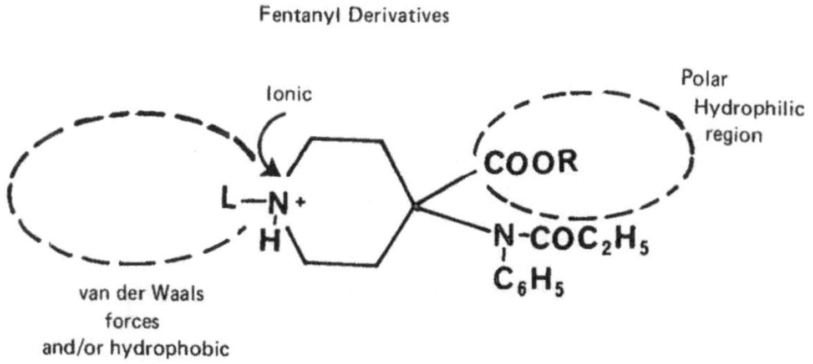

Fentanyl Derivatives

(20)

$$\log \frac{1}{C} = 1.907 + 2.970 \log MW_L - 0.184f_L^2 + 0.913f_L - 0.931f_R$$

$f_{L_o} = 2.48 \ (1.78-3.06)$

n	r	s
35	0.871	0.433

$F_{1,30} = 8.15$, significant at 99.5%

It is interesting to note that the ideal hydrophobicity as reflected by the f_{L_0} from the subset (n = 35) is quite similar to that from the complete set (n = 60), namely 2.48 vs. 2.25 (<u>21</u>).

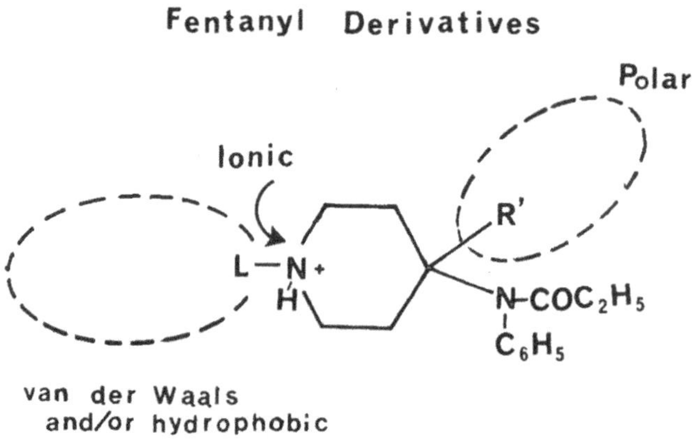

Fentanyl Derivatives

(<u>21</u>)

$$\log \frac{1}{C} = -0.399 + 3.300 \log MW_L - 0.230 f_L^2 + 1.114 f_L - 1.069 f_{R'}$$

f_{L_0} = 2.25 (1.50-2.63)

n	r	s
60	0.861	0.488

$F_{1,55}$ = 16.28, significant at 99.9%

One of these fentanyl derivatives, sufentanil citrate, has recently been approved by FDA as a class 1-B analgesic/anaesthetic (Sufenta—Janssen):

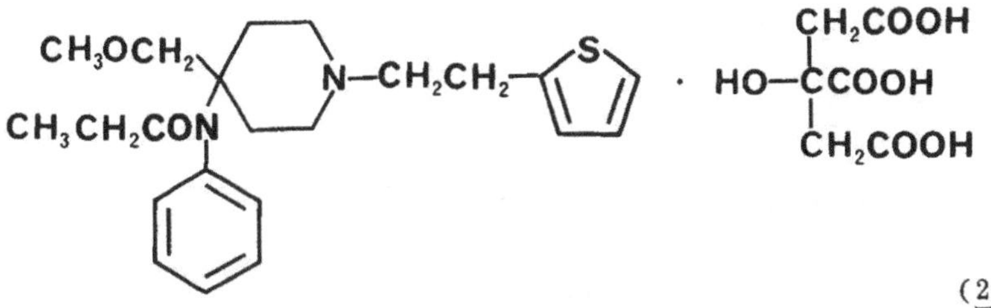

CH$_3$OCH$_2$ \
CH$_3$CH$_2$CON — N—CH$_2$CH$_2$—S

CH$_2$COOH \
HO—CCOOH \
CH$_2$COOH

(22)

Sufentanil citrate: *N*-[4-(methoxymethyl)-1-[2-(2-thienyl)ethyl]-4-piperidinyl]-*N*-phenylpropanamide-2-hydroxy-1,2,3-tricarboxylate (1:1) [Hussar, D. A. (1985). New drugs 1984, *Am. Pharm. NS25*, 50]

The structural similarities between some highly potent synthetic fentanyl analog (R 30490) and Met[5]-enkephalin, and between neuroleptics (benperidol, spiperone, flufenazine, R 1472, etc.) and the α-helix portions of β-lipotropin have been analyzed by Soudijn and Wijngaarden (70). Extensive research and impressive results on endorphins have been reported during the last few years (71), but no practical useful drug has yet been produced.

VII. QSAR OF ANTIHISTAMINES AND ANTICHOLINERGICS

The biosynthesis and metabolism of histamine as well as its interaction with H$_1$ and H$_2$ receptors are as follows:

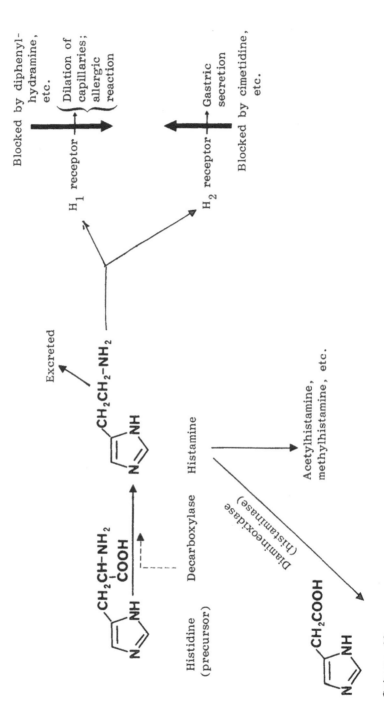

Scheme 18

From a series of 2-pyridylethylamines, the following QSARs have been derived for their antihistaminic (anti-H_1) and anti-cholinergic activities (6) (23):

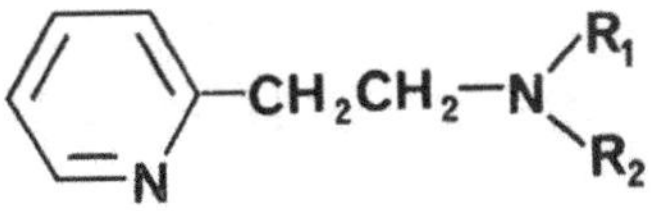

$$(23)$$

Antihistaminic activity in the guinea pig ileum:

	n	r	s	Significance level (%)
$pA_2 = 5.23 \log MV$ $-0.74n_H - 3.13$	29	0.91	0.56	> 99
$pD'_2 = 7.05 \log MV - 9.66$	29	0.93	0.41	> 99

Anticholinergic activity in the rat intestine:

	n	r	s	Significance level (%)
$pA_2 = 0.50 \log P + 3.10$	35	0.93	0.37	> 99
$pD'_2 = 0.65 \log P + 1.65$	34	0.94	0.46	> 99

In these equations MV is the molar volume (cm^3/mol) calculated from Bondi's table (72), and n_H is the number of hydrogen atoms for the protonated species ($n_H = 3$ for a primary amine).

The affinity constants (pA_2, log K) of 128 quaternary ammonium compounds belonging to several different series ($R-\overset{+}{N}\equiv$, $16 \times 8 = 128$) have been correlated linearly with the hydrophobicity (π_R), the dipole moment (μ_R), and the number of hydroxy groups (n_{OH}) of the side chain (73):

TABLE 25 Common Structural Features of Various Neurotransmitters

Neurotransmitter	Dipolar and/or H bonding	van der Waals and/or hydrophobic	Ionic
Acetylcholine (Ach)	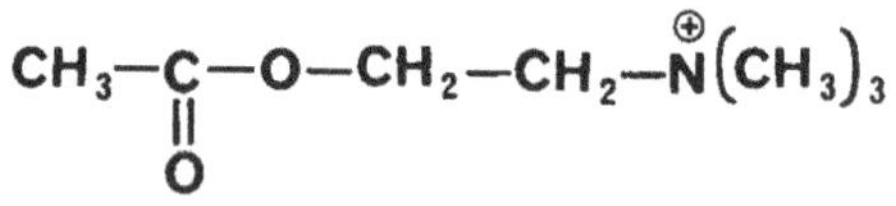 $CH_3-\overset{\displaystyle O}{\underset{\displaystyle \parallel}{C}}-O-CH_2-CH_2-\overset{\oplus}{N}(CH_3)_3$		
Norepinephrine	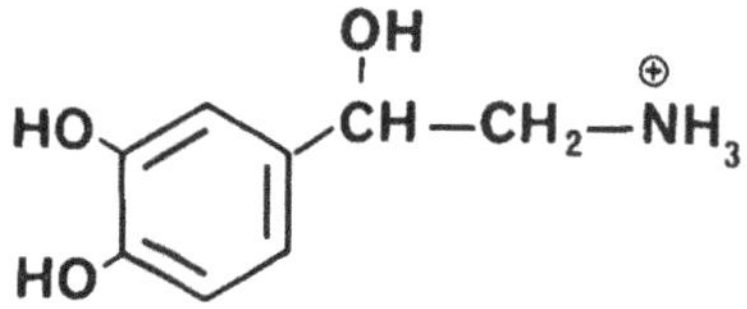		
Dopamine	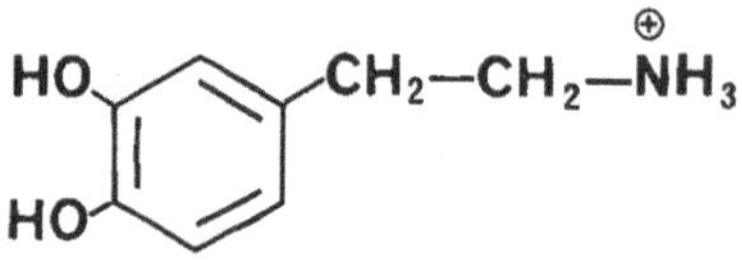		
Serotonin	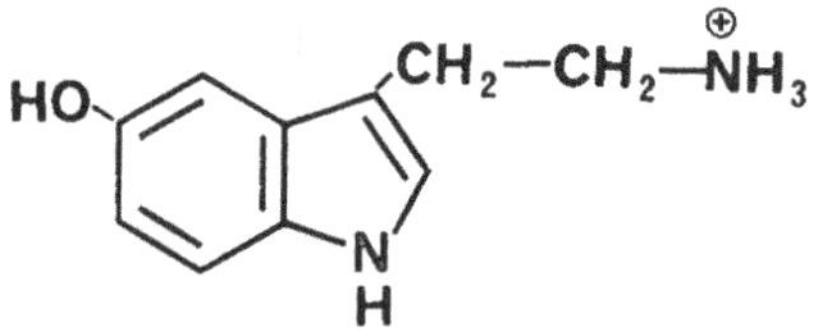		
Histamine	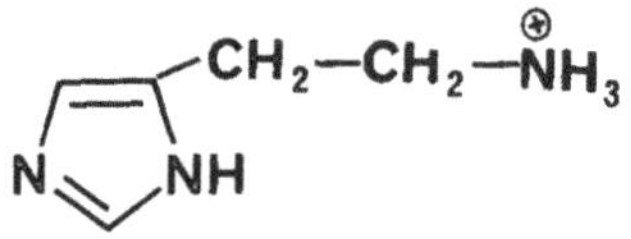		
γ-Aminobutyric acid	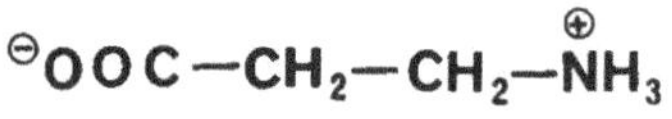 $^{\ominus}OOC-CH_2-CH_2-\overset{\oplus}{N}H_3$		

$$pA_2 = 0.78\pi_R - 0.35(\pi_{-N\equiv}^+)^2 - 0.17\pi_{-N\equiv}^+ + 0.74\mu_R + 2.31n_{OH} + 2.17$$

n	r	s
128	0.96	0.44

$$F_{1,122} = 6.98, \text{ significant at } 99\%$$

The dependence on the hydrophobicity constant of the quaternary ammonium head ($\pi_{-N\equiv}^+$) is shown to be parabolic. The ideal $(\pi_{-N\equiv}^+)_o$ value for maximum affinity is around -0.24, which corresponds to the value of $CH_3N(C_2H_5)_2$. It is not surprising, then, to find this group in several potent anticholinergic drugs, such as methantheline, oxyphenonium, and panthienate.

It is intriguing to see the common structural features among different neurotransmitters. They all have similar overall dimensions and functional groups to allow them to interact with their specific receptors by ionic, dipolar, and/or H bonding, van der Waals, and/or hydrophobic interactions (74). This is shown in Table 25.

This may explain why many antihistamines have anticholinergic activities, and vice versa. Drugs such as phenothiazines may bind to practically every receptor to some different extent. Although this may not be desirable from the therapeutical point of view, it does, nevertheless, widen the possibility for the discovery of "new drugs" by molecular modification, as shown by the numerous derivatives of antihistamines of the following general structures:

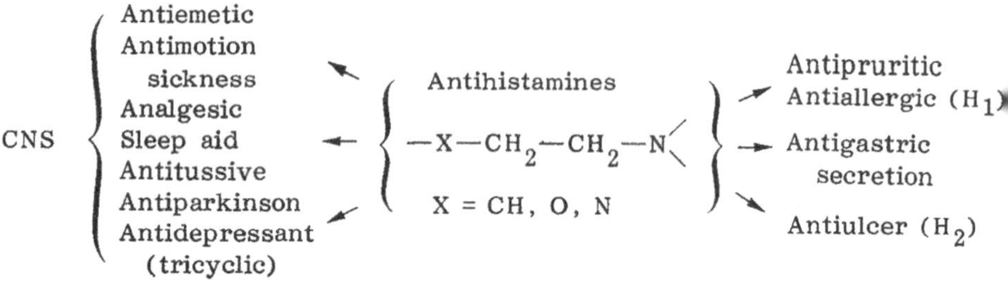
CNS
Antiemetic
Antimotion sickness
Analgesic
Sleep aid
Antitussive
Antiparkinson
Antidepressant (tricyclic)
Antihistamines
—X—CH2—CH2—N<
X = CH, O, N
Antipruritic
Antiallergic (H1)
Antigastric secretion
Antiulcer (H2)

A. H_2-Histamine Antagonists

Two compounds shown in the early 1970s to be selective inhibitors of H_2-histamine receptors are burimamide and metiamide. Both are thiourea derivatives with an imidazole connected to the dipolar thiourea group by a distance of four carbon atoms (24, 25):

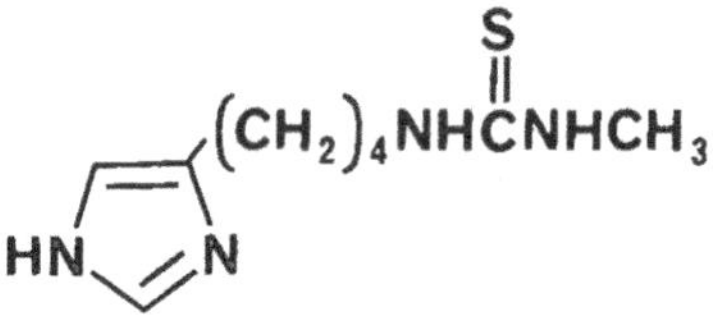
S
(CH2)4NHCNHCH3
HN
N

(24)

Burimamide — First drug introduced; only of historical interest

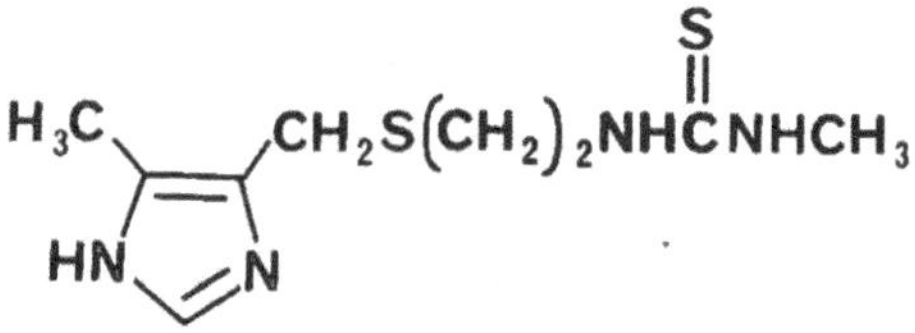
S
H3C
CH2S(CH2)2NHCNHCH3
HN
N

(25)

Metiamide — Good clinical effect but may cause bone marrow depression

Since then, cimetidine (Tagamet) has been marketed successfully for the treatment of duodenal ulcer (300 mg, p.o., q.i.d.) (75) (26). More recently, ranitidine (Zantac) has also been approved by the Food and Drug Administration (FDA) for similar uses (150 mg, b.i.d.) (76) (27):

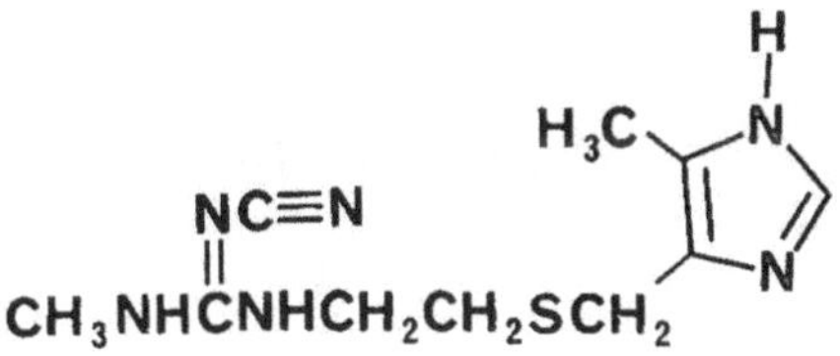

$$NC{\equiv}N$$

$$CH_3NHCNHCH_2CH_2SCH_2$$

H₃C — imidazole ring with N—H

(26)

Cimetidine

$$HC{-}NO_2$$

$$CH_2SCH_2CH_2N{-}C{-}NCH_3$$

$$H_2C{-}N(CH_3)_2$$

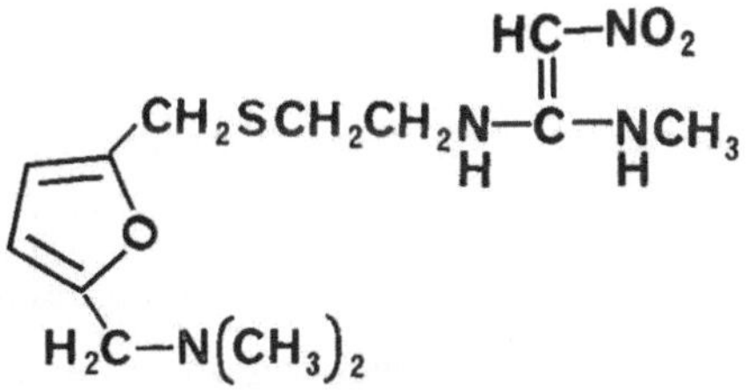

(27)

Ranitidine

The following overview of SAR shows the similarities and differences between H_1 and H_2 receptor antagonists and agonists (28, 29).

H_1 receptor:

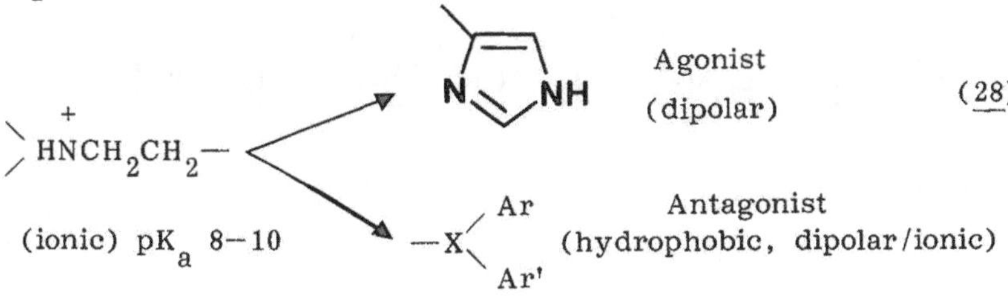

$$\overset{+}{>}HNCH_2CH_2{-}$$

(ionic) pK_a 8–10

Agonist (dipolar)   (28)

Antagonist (hydrophobic, dipolar/ionic)

$$-X\begin{smallmatrix} Ar \\ Ar' \end{smallmatrix}$$

H_2 receptor:

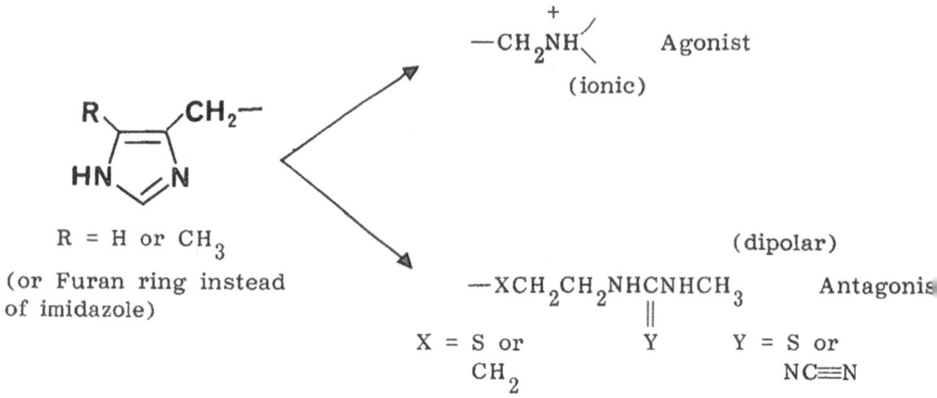

$R = H$ or CH_3
(or Furan ring instead
of imidazole)

$-CH_2NH\big\langle$ Agonist
(ionic)

(dipolar)

$-XCH_2CH_2NHCNHCH_3$ Antagonist

$X = S$ or Y $Y = S$ or
CH_2 $NC\equiv N$

($\underline{29}$)

Some important physicochemical properties of antihistamines
are shown in Table 26. It is apparent from this table that the
H_1 antagonists different from the agonists and H_2 antagonists
in both log P and percentages (% C_{aq}) remaining in aqueous
layer after equilibration with 1-octanol.

The dipolar nature of the cyanoguanidine (present in cimetidine) is shown as follows (Scheme 19):

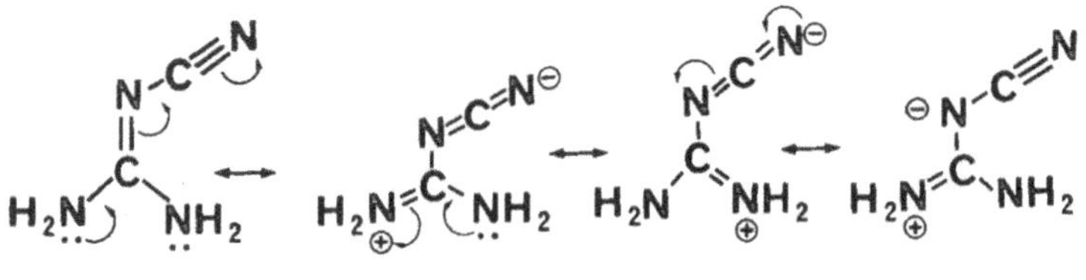

Scheme 19

TABLE 26 Partition Coefficient (log P) of Uncharged Forms of Histamine Agonists and Antagonists

	$P = \dfrac{C_{octanol}}{C_{aqueous\ buffer}}$	log P	$\%C_{aq} = \dfrac{100}{P+1}$
Histamine	0.2	−0.7	83
Metiamide sulfoxide	0.2	−0.7	83
Imidazole	0.5	−0.3	67
4(5)-Methylimidazole	1.5	0.18	40
Thiaburimamide	1.4	0.16	42
Burimamide	2.5	0.40	29
Cimetidine	2.5	0.40	29
Metiamide	3.2	0.51	24
Ranitidine	5.0	0.7[a]	17
Methylburimamide	7.1	0.85	12
Mepyramine	700	2.85	0.14
Diphenhydramine	2500	3.40	0.04
Triprolidine	8300	3.92	0.012

(log P: low ⟷ high; $\%C_{aq}$: significant ⟷ negligible)

[a]Calculated value: A. Leo, Pomona College, personal communication.
Source: Adapted from Ref. 77.

In addition to resonance, both ionization and tautomerism can take place, depending on the pH of the solution or the environment (77) (Scheme 20).

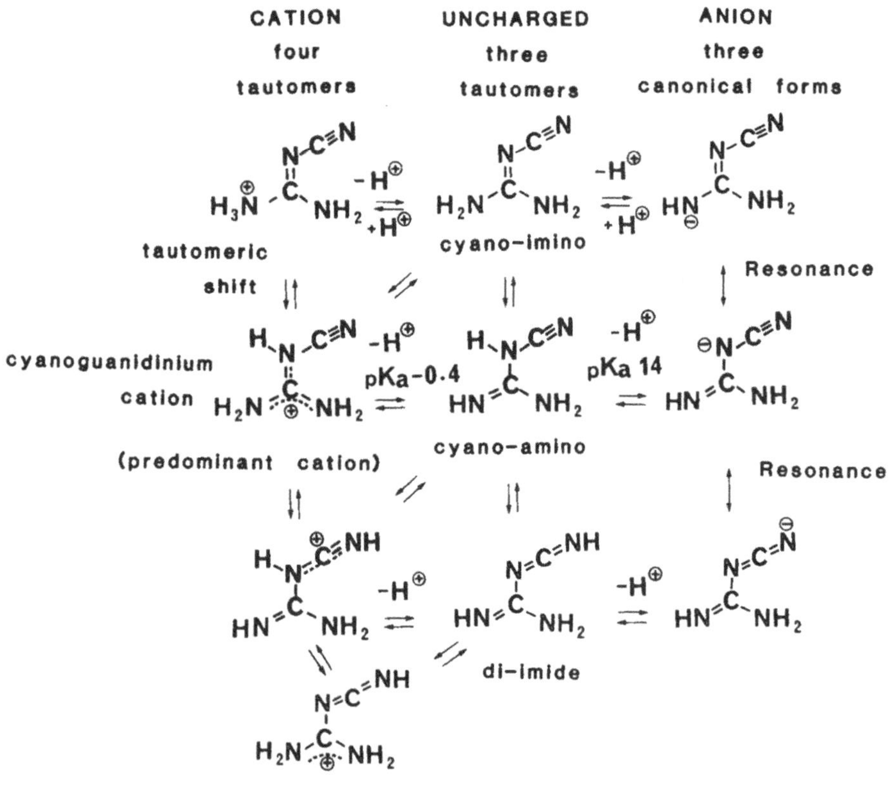

Scheme 20

In the case of ranitidine the $-N-C\equiv N$ group has been substituted by the $\equiv CH-NO_2$ group in addition to replacement of the imidazole ring by a furan ring.

VIII. QSAR OF STEREOISOMERS

A. Optical Isomers

Most structure-activity relationship studies on stereoisomers are either qualitative or semiquantitative in nature. Among the empirical generalizations reported, Pfeiffer's rule (78) appears to be valid in many cases, although there are some exceptions (79). The principle of Pfeiffer's rule states that the activity ratio of the isomeric pair of a more active compound is higher than that of a less active compound. This is shown in Figure 18.

Optical isomers are usually not included in quantitative structure-activity correlation studies, mainly because (a) 1-

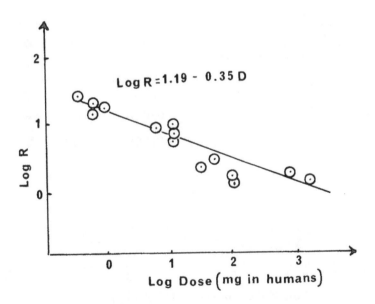

Figure 18 Pfeiffer's plot of the log activity ratio vs. log dose.

octanol-water or other solvent systems used in determining
the partition coefficeints cannot differentiate the partition be-
haviors of enantiomorphs, (b) not many sets of quantitative data
on stereoisomers are available for QSAR analysis, and (c) quantitative data on stereoisomers are more difficult to obtain (con-
tamination of the less active isomer by the active one will significantly change the activity of the less active isomer).

Although some investigators have used an indicator variable
(or dummy parameter, say 1 for the R isomers and 0 for the S
isomers) to take into account the different stereoisomers, this
author feels that such an approach is a compromise in opposition
to Pfeiffer's rule.

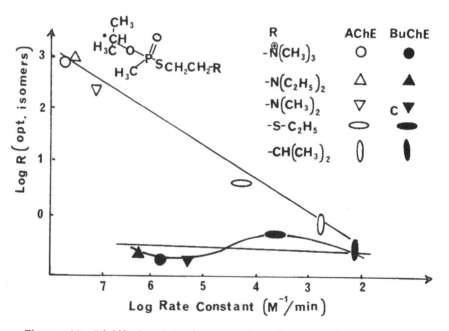

Figure 19 Pfeiffer's plot of organophosphates against acetylcholinesterase. (From Ref. 79.)

A stereo-discriminatory phenomenon exists in homologs and congeners as well as miscellaneous groups of compounds originally reported by Pfeiffer. The following illustration taken from Ariëns' book clearly shows that for the inhibitors of true cholinesterase by pairs of stereoisomers of a series of organophosphates, Pfeiffer's rule is obeyed, while for the pseudo-cholinesterase much less selectivity is observed (79) (Figure 19).

Ariëns has shown further that if two chiral centers are involved, for the more critical center in receptor binding a larger ratio will be observed for the isomer pair than for the less critical center. Scheme 21 suggests that the α-carbon to the

Ratio

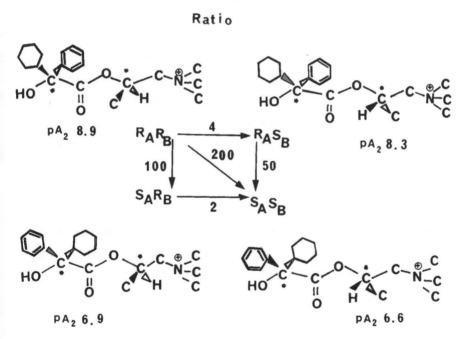

Scheme 21 Affinity constants of anticholinergics on rat jejunum.

carbonyl group is more critical than the β-carbon to the quaternary ammonium head in drug-receptor interaction (79).

Keeping these findings in mind, we have developed a mathematical model for QSAR study when *two* substituents at a chiral center are changed simultaneously. By treating the physicochemical properties of the *two chiral substituents* at the optically active center (see Tables 27 and 28) as *independent variables*, the variation in the biological activities of many pairs of enantiomorphs can be quantitatively correlated with the molecular structures. Because of the availability of the biological data on a large number of optically active aryloxypropionic acid derivatives, these are used as an example in regression analysis (80).

Since all enzymes and/or receptor proteins are made of naturally occurring optically active L-amino acids (except glycine), any drug-receptor interaction of high specificity should be considered as a drug molecule occupying a chiral environment. Consequently, the two substituents of a highly active R isomer should contribute to the binding energy quite differently from its enantiomorph (the S isomer) (Scheme 22).

The auxin activity of 56 optically active aryloxypropionic acids and structurally related derivatives are taken from the compilation of Jönsson (81) and Åberg (82) (Table 27), where C_{50} is the concentration of the substance that reduces growth to 50% of that of control in the flax root test. Compounds without an oxygen or other heteroatom connecting the aromatic ring and the propionic acid moiety (e.g., indolepropionic acids) are excluded in the study, to avoid too many structural variables for just a few additional compounds.

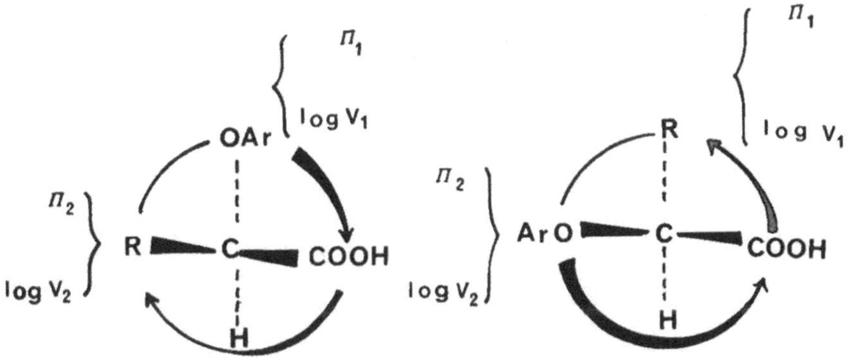

D(+) isomer = R configuration   L(−) isomer = S configuration
(few exceptions)

Exceptions: (+) isomer = L series:
2-Iodophenoxypropionic acid
α-Naphthoxypropionic acid
1-Naphthoxy-n-butyric acid

Scheme 22

After trying many different combinations of various parameters, equations 1 to 7 in Table 28 appear to be satisfactory in
describing the auxin activities of the compounds examined.
From these equations one can see that by using a quadratic
equation of the π constant and log volume terms for both R_1 and
R_2 plus the σ constant of the aromatic ring, a correlation coefficient (r) of 0.8 is obtained for all 56 compounds examined
(equation 1). After deleting nine outliers, equation 7 appears
to be the most representative equation. About 81% ($r^2 = 0.81$)
of the variance can be explained for the 47 compounds included
in equation 7.

TABLE 27 Physicochemical Constants and Biological Data for Aryloxypropionic Acid Derivatives Inhibiting Flax Root Growth[a]

$$R_2 - \overset{R_1}{\underset{H}{C}} - COOH$$

Eq. no.	log 1/C$_{50}$ Obsd.[b]	log 1/C$_{50}$ Calcd.[c]	log V$_{R_1}$	π_{R_1}	π_{R_2}	log V$_{R_2}$	σ_{Ar}	Configuration	R$_1$	R$_2$
1	5.89	6.22	1.69	1.57	0.56	1.14	0.00	R(+)	Ph—O—	Me—
1'	3.96	4.18	1.14	0.56	1.57	1.69	0.00	S(−)	Me—	Ph—O—
2	7.48	6.49	1.82	3.03	0.56	1.14	0.45	R(+)	2,4-DiCl—Ph—O—	Me—
2'	4.89	4.99	1.14	0.56	3.03	1.82	0.46	S(−)	Me—	2,4-DiCl—Ph—O—
3	7.55	6.65	1.82	3.03	0.56	1.14	0.60	R(+)	3,4-DiCl—Ph—O—	Me—
3'	5.46	5.15	1.14	0.56	3.03	1.82	0.60	S(−)	Me—	3,4-DiCl—Ph—O—
4	5.00	5.09	1.14	0.56	3.79	1.87	0.83	S(−)	Me—	2,4,5-TriCl—Ph—O—
5	7.68	(5.94)[d]	1.83	3.11	0.56	1.14	0.06	R(+)	4-Cl-2-Me—Ph—O—	Me—
5'	4.74	4.48	1.14	0.56	3.11	1.83	0.06	S(−)	Me—	4-Cl-2-Me—Ph—O—
6	5.85	5.43	1.69	1.57	1.02	1.38	0.00	R(+)	Ph—O—	Et—
6'	4.05	4.32	1.38	1.02	1.57	1.69	0.00	S(−)	Et—	Ph—O—
7	7.55	(6.03)[d]	1.82	3.03	1.02	1.38	0.46	R(+)	2,4-DiCl—Ph—O—	Et—
7'	4.85	5.13	1.38	1.02	3.03	1.82	0.46	S(−)	Et—	2,4-DiCl—Ph—O—

No.										
8	7.49	(5.93)d	1.87	3.79	0.56	1.14	0.83	R(+)	2,4,5-TriCl—Ph—O—	Me—
9	3.52	(5.21)d	1.69	1.57	2.04	1.65	0.00	R(+)	Ph—O—	n-Bu—
9'	3.52	4.51	1.65	2.04	1.57	1.69	0.00	S(−)	n-Bu—	Ph—O—
10	7.70	(5.99)d	1.87	2.84	0.56	1.14	0.04	R(+)	2-Naph—O—	Me—
10'	4.09	4.10	1.14	0.56	2.84	1.87	0.04	S(−)	Me—	2-Naph—O—
11	4.84	(5.99)d	1.87	2.84	0.56	1.14	0.04	R(−)	1-Naph—O—	Me—
11'	4.64	4.10	1.14	0.56	2.84	1.87	0.04	S(+)	Me—	1-Naph—O—
12	4.33	4.24	1.38	1.02	2.84	1.87	0.04	S(+)	Et—	1-Naph—O—
13	7.00	(5.11)d	1.87	2.84	1.02	1.38	0.04	R(+)	2-Naph—O—	Et—
13'	4.14	4.24	1.38	1.02	2.84	1.87	0.04	S(−)	Et—	2-Naph—O—
14	4.92	5.43	1.92	3.60	0.56	1.14	0.27	R(+)	1-Cl-2-Naph—O—	Me—
14'	4.21	4.13	1.14	0.56	3.60	1.92	0.27	S(−)	Me—	1-Cl-2-Naph—O—
15	5.55	4.73	1.87	2.04	2.84	1.65	0.04	R(+)	2-Naph—O—	n-Bu—
15'	4.30	4.43	1.65	2.84	2.04	1.87	0.04	S(−)	n-Bu—	2-Naph—O—
16	5.66	5.43	1.73	0.98	0.56	1.14	0.00	R(+)	Ph—NH—	Me—
16'	3.52	3.00	1.14	0.56	0.98	1.73	0.00	S(−)	Me—	Ph—NH—
17	6.74	(5.78)d	1.91	3.56	0.56	1.14	0.04	R(+)	2-Naph—S—	Me—
17'	4.30	3.94	1.14	0.56	3.56	1.91	0.04	S(−)	Me—	2-Naph—S—
18	6.0e	6.35	1.71	1.72	0.56	1.14	0.06	R(+)	4-F—Ph—O—	Me—
18'	4.0e	4.30	1.14	0.56	1.72	1.71	0.06	S(−)	Me—	4-F—Ph—O—
19	6.5e	6.62	1.76	2.27	0.56	1.14	0.23	R(+)	4-Cl-Ph—O—	Me—
19'	4.4e	4.72	1.14	0.56	2.27	1.76	0.23	S(−)	Me—	4-Cl-Ph—O—

TABLE 27 (Continued).

Eq. no.	log 1/C_{50} Obsd.[b]	Calcd.[c]	log V_{R_1}	π_{R_1}	π_{R_2}	log V_{R_2}	σ_{Ar}	Configuration	R_1	R_2
20	6.6[e]	6.55	1.78	2.59	0.56	1.14	0.23	R(+)	4-Br—Ph—O—	Me—
20'	4.5[e]	4.81	1.14	0.56	2.59	1.78	0.23	S(−)	Me—	4-Br—Ph—O—
21	7.4[e]	6.56	1.81	2.72	0.56	1.14	0.35	R(+)	3-I—Ph—O—	Me—
21'	5.1[e]	4.81	1.14	0.56	2.72	1.81	0.35	S(−)	Me—	3-I—Ph—O—
22	6.2[e]	6.32	1.81	2.83	0.56	1.14	0.18	R(+)	4-I—Ph—O—	Me—
23	5.9[e]	6.10	1.77	2.09	0.56	1.14	−0.17	R(+)	4-Me—Ph—O—	Me—
24	7.3[e]	6.61	1.82	3.09	0.56	1.14	0.60	R(+)	2,5-DiCl—Ph—O—	Me—
24'	5.0[e]	5.17	1.14	0.56	3.09	1.82	0.60	S(−)	Me—	2,5-DiCl—Ph—O—
25	6.5[e]	6.61	1.82	3.09	0.56	1.14	0.60	R(+)	2,3-DiCl—Ph—O—	Me—
25'	5.0[e]	5.17	1.14	0.56	3.09	1.82	0.60	S(−)	Me—	2,3-DiCl—Ph—O—

26	5.4[e]	5.01	1.14	0.56	3.09	1.82	0.46	S(−)	Me—	Me—	2,6-DiCl—Ph—O—
27	5.4[e]	(6.77)[d]	1.82	3.09	0.56	1.14	0.74	R(+)	3,5-DiCl—Ph—O—	Me—
27'	4.8[e]	5.33	1.14	0.56	3.09	1.82	0.74	S(−)	Me—	3,5-DiCl—Ph—O—
28	5.1[e]	5.77	1.87	3.79	0.56	1.14	0.69	R(+)	2,4,6-Tricl—Ph—O—	Me—
28'	4.6[e]	4.93	1.14	0.56	3.79	1.87	0.69	S(−)	Me—	2,4,6-TriCl—Ph—O—
29	4.7[e]	4.59	1.92	4.55	0.56	1.14	1.06	R(+)	2,4,5,6-TetraCl—Ph—O—	Me—
29'	4.6[e]	4.61	1.14	0.56	4.55	1.92	1.06	S(−)	Me—	2,4,5,6-TetraCl—Ph—O—
30	6.0[e]	6.43	1.81	2.49	0.56	1.14	0.18	R(−)	2-I—Ph—O—	Me—
30'	5.0[e]	4.48	1.14	0.56	2.49	1.81	0.18	S(+)	Me—	2-I—Ph—O—
31	5.6[e]	6.45	1.82	3.09	0.56	1.14	0.46	R(+)	2,6-DiCl—Ph—O—	Me—
32	4.3	4.45	1.99	3.93	0.56	1.14	0.01	R(+)	2-i-Pr-4-Cl-5-Me—Ph—O—	Me—

[a] R_1 = aryloxy and R_2 = alkyl for the R(+) isomers, with the notable exceptions of 2-iodophenoxypropionic acid, α-naphthoxypropionic acid, and 1-naphthoxy-n-butyric acid; the (+) isomer belongs to the S series.
[b] From Ref. 81 unless stated otherwise.
[c] Calculated from equation 7.
[d] Not included in the derivation of equation 7.
[e] From Ref. 82.

TABLE 28 Equations Correlating Auxin Activity with Physicochemical Constants of Optical Isomers of Phenoxypropionic Acid Derivatives

$$\log 1/C_{50} = -k_1(\pi_1)^2 + k_2\pi_1 + k_3 \log V_1 + K_4\sigma Ar - k_5(\pi_2')^2 + k_6\pi_2 + k_7 \log V_2 + k_8$$

Eq. no.	k_1	k_2	k_3	k_4	k_5	k_6	k_7	k_8	n	r	s
1	0.569	3.224	-4.260	0.583	0.289	2.190	-6.858	16.365	56	0.802	0.767
2	0.567	3.423	-5.085		0.273	2.306	-7.543	18.186	56	0.796	0.769
3	0.586	3.342	-5.074	0.496	0.286	2.324	-8.026	19.004	52	0.864	0.642
4	0.585	3.270	-4.965	0.439	0.288	2.343	-7.994	18.846	50	0.874	0.585
5	0.581	3.381	-5.506		0.275	2.428	-8.516	20.155	50	0.870	0.585
6	0.549	2.932	-4.210	0.688	0.295	2.278	-7.529	17.469	49	0.878	0.552
7	0.534	2.514	-2.619	1.151	0.356	2.442	-6.886	14.606	47[a]	0.902	0.484

[a] The most representative equation.

From the parabolic equation one can derive the ideal lipophilic requirements (π_o) of the receptor for R_1 and R_2. These are shown as follows:

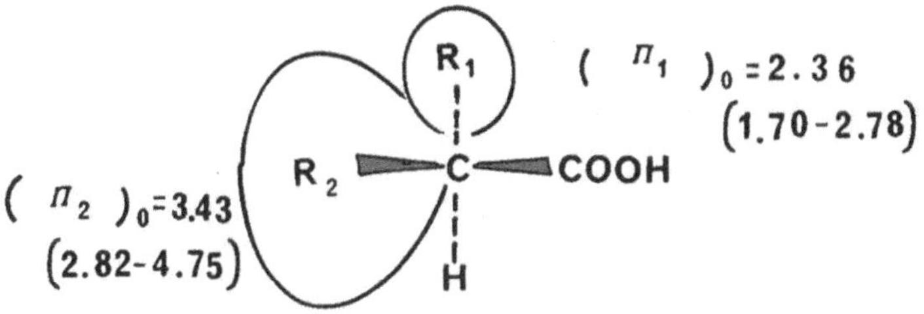

R1
(π1)0 = 2.36
(1.70-2.78)
R2
C
COOH
(π2)0 = 3.43
(2.82-4.75)
H

For both R (D) and S (L) isomers

For R (D) isomers alond, n = 29 $(\pi_1)_o$ = 2.51 (0.91–4.51)

S (L) isomers alone, n = 27 $(\pi_2)_o$ = 2.82 (2.06–3.76)

Specific examples of stereoselectivity in drug and xenobiotic metabolism, drug distribution, and receptor binding have been reviewed (83).

B. Cis-Trans Isomers

We have used the same approach in the quantitative structure-activity correlation of geometric isomers acting as antihistamines (84) (see also Table 29).

H_1 receptor of the guinea pig ileum with antihistamines

R′(trans)
H
C=C
CH2
N
R(cis)

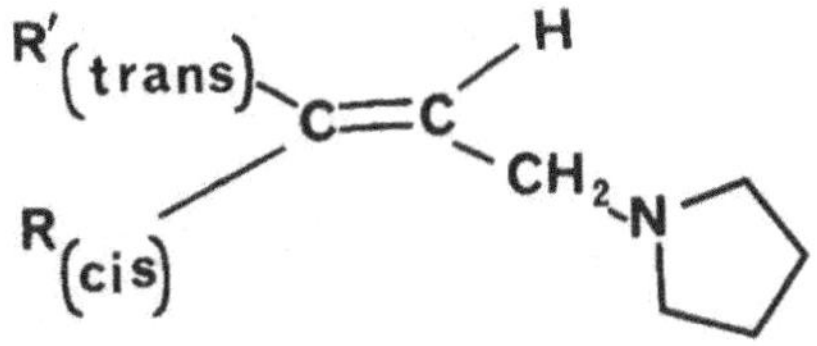

Geometric isomers of 1,1-diaryl-3-aminopro-1-enes- and 1,2-diaryl-4-aminobut-2-enes

TABLE 29 Affinity Constants for Histamine Receptors of the Guinea Pig Ileum and the Physico-chemical Constants Used in the Regression Analysis

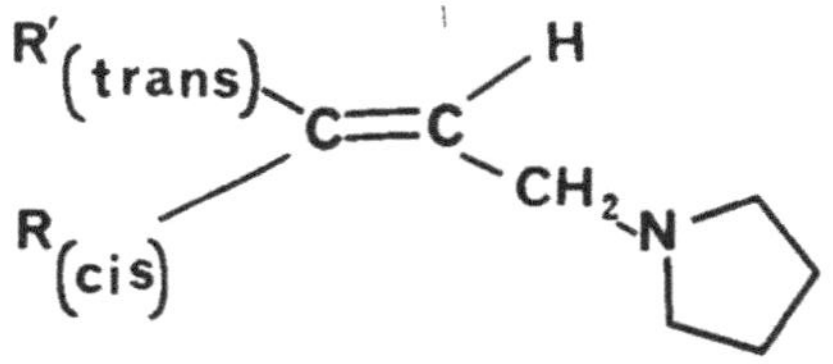

Compd. no.	log K_B or pA_2 Obsd.	Calcd.[a]	log V_R	π_R	$\pi_{R'}$	R	R'
1	8.66[b] (8.35 ± 0.10)[c,d,f]	8.71	1.66	1.96	0.32	Ph	α-Pyridyl
2	7.17 ± 0.10[c,d,f] (7.69)[b]	6.77	1.64	0.32	1.96	α-Pyridyl	Ph
3	8.61[b] (8.55 ± 0.09)[c,d,g]	9.01	1.74	2.67	0.32	p-Cl—Ph	α-Pyridyl
4	5.97 ± 0.10[c,d,h] (7.78)[b]	5.88	1.64	0.32	2.67	α-Pyridyl	p-Cl—Ph
5	9.95[b] (8.15 ± 0.10)[c,d,h] (9.0)[e]	9.53	1.75	2.52	0.32	p-Toyl	α-Pyridyl

6	5.60 ± 0.10[c,d,g] (6.88)[b]	6.10	1.64	0.32	2.52	α-Pyridyl	p-Toyl
7	10.34[b]	9.87	1.81	2.67	1.96	p-Cl—Ph—CH$_2$—	Ph
8	7.97[b]	7.38	1.66	1.96	2.67	Ph	p-Cl—Ph—CH$_2$—
9	8.15[b]	8.27	1.66	1.96	1.96	Ph	Ph
10	7.66[b]	8.42	1.75	2.52	2.52	p-Tolyl	p-Tolyl
11	8.00[b]	7.68	1.74	2.67	2.67	p-Cl—Ph—	p-Cl—Ph—
12	8.45[b]	8.57	1.74	2.67	1.96	p-Cl—Ph—	p-Cl—Ph—
13	7.30[b]	7.38	1.66	1.96	2.76	Ph	p-Cl—Ph
14	9.64[b]	9.90	1.75	2.01	1.96	Ph—CH$_2$—	Ph

[a]Calculated from equation 8.
[b]From Ison et al. (85).
[c]The pA_2 value was from this study; the sample was obtained from J. Billinghurst and A. Green of the Wellcome Research Labs., Beckenham, Kent, England.
[d]Standard error from 12 to 25 measurements on specimens from 2 to 5 animals.
[e]From Ison and Casy (91).
[f]The oxalate was used.
[g]The HCl salt was used.
[h]The HCl monohydrate was used.

The SAR of a series of conformationally restricted geometric isomers of 1,1-diaryl-3-aminoprop-1-enes and 1,2-diaryl-4-aminobut-2-enes is shown in Table 30. The affinity constants (log K_B or pA_2), defined as the negative logarithm of the antagonist concentration [B] which requires double the amount of agonist for canceling the effect of antagonist, are either from the paper of Ison et al. (85) or experimentally determined (86, 87) using the standard method of cumulative concentration-effect curves.

From equations 1 to 5 it is apparent that bulky and hydrophobic group R at the position cis to the pyrrolidine group will enhance the affinity constant; the contrary is true for the trans substituent R'. The dependence on π_R and $\pi_{R'}$ appears to be parabolic, while the dependence on log V_R appears to be linear for the 14 compounds examined. The simultaneous additions of the quadratic and linear terms of π_R and $\pi_{R'}$ are statistically significant judging from the F test, although the stepwise additions of the $(\pi_R)^2$ and $(\pi_{R'})^2$ terms are not significant.

Equation 8 is considered as the "best" correlation for the data available. Six terms are included in this equation; however, only three of them are independent variables (log V_R, π_R, and $\pi_{R'}$). It may be possible that by including more compounds with different R' and by using more refined steric parameters, better correlations may be obtained. It is encouraging to see that a wide range of the affinity constants of the geometric isomers of antihistamines can be described quantitatively in physicochemical terms.

This gives one of the possible molecular bases of Pfeiffer's rule applied to geometric isomers (89, 90); that is, the isomeric

TABLE 30 The SAR of a Series of Geometric Isomers of 3-Aminoprop-1-enes and 4-Aminobut-2-enes

Eq. no.	n	r	s
1 $\log K_B = -0.815 \pi_{R'} + 9.65$	14	0.54	1.19
2 $\log K_B = -14.50 \log V_{R'} + 32.70$	14	0.60	1.14
3 $\log K_B = 1.14 \pi_R + 5.95$	14	0.76	0.92
4 $\log K_B = 18.58 \log V_R - 23.53$	14	0.77	0.90
5 $\log K_B = 11.13 \log V_R + 0.60 \pi_R - 11.97$	14	0.81	0.87
6 $\log K_B = 21.82 \log V_R - 1.10(\pi_R)^2 + 3.27 \pi_R - 30.48$	14	0.88	0.74
7 $\log K_B = 22.62 \log V_R - 1.15(\pi_R)^2 + 3.20 \pi_R - 0.54 \pi_{R'} - 30.46$	14	0.94	0.54
8 $\log K_B = 18.63 \log V_R - 1.01(\pi_R)^2 + 2.98 \pi_R - 0.42(\pi_{R'})^2 + 0.70 \pi_{R'} - 24.38$	14	0.96	0.51

ratio of the activity or $\Delta \log K_B$ increases with increasing activity of the more active isomer. If one realizes that the $(\pi_R)_0$ is different from $(\pi_{R'})_0$ and the dependence on $\log V_R$ is positive while the dependence on $\log V_{R'}$ is negative, it becomes understandable why the $\Delta \log K_B$ of a highly active member with its geometrical isomer will be greater than that of a moderately active pair. In other words, because of the different structural requirements for the cis and trans substituents, the reversal of an ideal configuration will lower the affinity drastically, whereas the reversal of a mediocre configuration will make little or no difference in the affinity to the receptor.

$$(\pi_R)_0 = 1.40 \ (1.12-2.03) \qquad \text{from equation 7}$$

$$\left.\begin{array}{l} (\pi_R)_0 = 1.48 \ (1.17-2.88) \\[6pt] (\pi_{R'})_0 = 0.83 \pm \infty \end{array}\right\} \quad \text{from equation 8}$$

The optimum hydrophobic character of the cis group $(\pi_R)_0$ for maximum affinity is 1.48, with a sharp 95% confidence interval of 1.17 to 2.88. The $(\pi_{R'})_0$ is only 0.83, but the 95% confidence interval cannot be defined.

This clearly indicates that the hydrophobic nature and the bulk tolerance of the binding site for R are different from those for R'. The positive dependence on $\log V_R$ suggests that the blocking effect of the antihistamine molecule against the access of the histamine to the anionic site (the binding site for the protonated tertiary amino group) from the cis position is proportional to the bulk of the substituent R. The addition of $\log V_{R'}$ does not improve the correlation significantly. This is quite logical, since the agonist (histamine) probably can approach the anionic site from the nearby cis position more effectively than from the trans position.

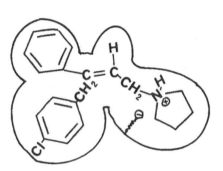 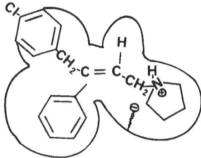

Cis isomer
Perfect fit
$\log K_B$ = 10.34

Trans isomer $+ / ^H$
Good fit for the $-N\equiv$ group
Poor fit for the $Cl-Ph-CH_2-$
and Ph groups
$\log K_B$ = 7.97

Since the largest R group examined is $p-Cl-Ph-CH_2-$ with a log V value of 1.81, the linear dependence on log V_R should not be taken as an indication of unlimited bulk tolerance. Exactly how far the linearity holds remains to be studied. It is interesting to note that a positive dependence on log V was also observed for the side-chain substituents of a series of 2-β-aminoethylpyridine derivatives (6).

C. Conformationally Restricted Compounds

Another difficult problem in QSAR is that of quantitatively describing different conformations or degrees of rotation around a certain chemical bond. The example is the relationship between estrogenic activity and the angle of twist of the two benzene rings in a series of 4,4'-dimethoxystilbenes.

Scheme 23 shows the general structure of a series of 4,4'-dimethoxystilbenes reported by Laarhoven et al. (92). Depending on the number of methyl or ethyl groups at the α, α', 2,2', and 6ˉ,6 positions, the angle of twist ranges from 0 to 90° for the unsubstituted stilbene to the tetramethyl derivative. The estrogenic activity as measured by cornification in vaginal smears has been found to be highly dependent on the angle of twist (Table 31). I choose to use $\sin\phi$ as the parameter in regression since it ranges from 0.0 to 1.0 for the compounds examined.

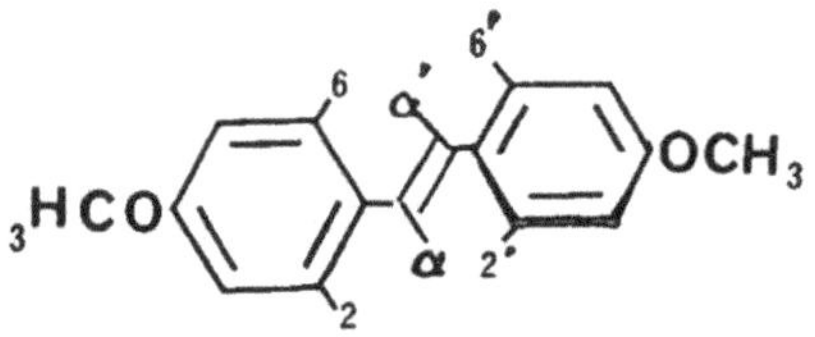

Angle of twist ϕ

Scheme 23

The single-parameter equation of $\sin\phi$ gives a highly significant correlation ($F_{1,10} = 42.3$). Addition of log P (calcd) does not improve the correlation significantly ($F_{1,9} = 1.56$). However, the addition of the $(\log P)^2$ and log P terms simultaneously gives the last equation, which is significant at the 95% level ($F_{2,8} = 5.67$, $F_{2,8_{0.05}} = 4.46$). This is a reversed parabola with a log P_0 for minimum activity around 5.97 (5.63–7.11). Whether this is due to competitive protein binding or other factors remains to be studied. Reversed parabolic equations have been found in the distribution of antibiotics from plasma to prostatic fluid (93). The positive dependence on $\sin\phi$ suggests that steric hindrance to coplanarity is essential to high activity in the 4,4'-dimethoxystilbenes. Whether this is due to

TABLE 31 The SAR of a Series of 4,4'-Dimethoxystilbenes

Estrogenic activity in mice	n	r	s
$\log 1/D = 1.023 \log P + 0.299$	12	0.57	1.01
$= -3.089\sigma + 3.334$	12	0.58	1.01
$= -0.407 (\log P)^2 + 5.520 \log P - 11.965$	12	0.59	1.05
$= 3.661 \sin\phi + 3.730$	12	0.899	0.54
$= 4.456 \sin\phi - 0.462 \log P + 5.833$	12	0.915	0.52
($F_{1,9} = 1.56$)			
$= 5.636 \sin\phi + 0.958(\log P)^2 - 11.435 \log P + 36.094$	12	0.960	0.39
($F_{2,8} = 5.67$; $\alpha < 0.05$)			

$\log P_0$ for minimum activity" 5.97 (5.63—7.11)

bioactivation, transport, or drug-receptor interaction remains to be studied.

An extensive compilation of the properties and structure-activity relationships of both agonists and antagonists of adrenergic and cholinergic receptors has recently been published by Szász (94).

Careful examination of the pharmacological profiles of stereo-isomers can result in better therapeutic agents. For example, Liu and Chiou (95) have discovered that the d isomer of timolol suppressed aqueous humor formation as effectively as l-timolol. In a randomized, double-masked, single-drop study of the effects of d-timolol and placebo on intraocular pressure in 34 patients with ocular hypertension, Keates and Stone (96) reported that d-timolol can lower intraocular pressure without inducing β-adrenergic blocking side effects. As a result a use patent for d-timolol was granted to Chiou and Liu in 1985 by the U.S. Patent Office.

REFERENCES

1. Litchfield, J. T., and Fertig, J. W. (1941). On a graphic solution of the dosage-effect curve. *Bull. Johns Hopkins Hosp. 69*: 276–286.

2. Miller, L. C., and Tainter, M. L. (1944). Estimation of the ED_{50} and its error by means of logarithmic-probit paper. *Proc. Soc. Exp. Biol. Med. 57*: 261–264.

3. Turner, R. A. (1965). *Screening Methods in Pharmacology.* Academic Press, New York, pp. 60–62.

4. Diem, K., Ed. (1962). *Documenta Geigy Scientific Tables,* 6th ed. Geigy Pharmaceuticals, Ardsley, N.Y., pp. 54–55.

5. Finney, D. J. (1971). *Probit Analysis,* 3rd ed. Cambridge University Press, London.

6. Van den Brink, F. G., and Lien, E. J. (1978). In, M. R. e Silva, Ed., *Handbook of Experimental Pharmacology*, Vol. XVIII/2. Springer-Verlag, Berlin, pp. 333–367.

7. Van den Brink, F. G. (1977). In, J. M. van Rossum, Ed., *Handbook of Experimental Pharmacology*, Vol. 47. Springer-Verlag, Berlin, pp. 169–254.

8. Hammett, L. P. (1940). *Physical Organic Chemistry*. McGraw-Hill, New York.

9. Lien, E. J. (1985). Structure, properties and disposition of drugs, in E. M. Jucker, Ed., *Progress in Drug Research*, Vol. 29, pp. 67–95.

10. Wells, P. R. (1968). *Linear Free Energy Relationships*. Academic Press, New York.

11. Gould, E. S. (1959). *Mechanism and Structure in Organic Chemistry*. Holt, Rinehart and Winston, New York.

12. Leffler, J. E., and Grunwald, E. (1963). *Rates and Equilibria of Organic Reactions*. Wiley, New York.

13. Hancock, C. K., and Falls, C. P. (1961). A Hammett-Taft polar-steric equation for the saponification rates of m- and p-substituted alkylbenzoates. *J. Am. Chem. Soc. 83*: 4214–4216.

14. Hansch, C., and Lien, E. J. (1968). An analysis of the structure-activity relationship in the adrenergic blocking activity of the β-haloalkylamines. *Biochem. Pharmacol. 17*: 713.

15. Saunders, L. (1966). *Principle of Physical Chemistry for Biology and Pharmacy*. Oxford University Press, London.

16. Cram, D. J., and Hammond, G. S. (1964). *Organic Chemistry*, 2nd ed. McGraw-Hill, New York, p. 216.

17. Brown, H. C., and Kanner, B. (1953). 2,6-Di-butyl-pyridine—an unusual pyridine. *J. Am. Chem. Soc. 75*: 3865.

18. Rappoport, Z. (1974). *CRC Handbook of Tables for Organic Compound Identification*, 3rd ed. Chemical Rubber Company, Cleveland, Ohio.

19. Windholz, M., Ed. (1983). *The Merck Index*. Merck Sharp & Dohme Research Laboratories, Rahway, N.J.

20. Newton, D. W., and Kluza, R. N. (1978). pK_a values of medicinal compounds in pharmacy practice. *Drug Intell. Clin. Pharm. 12*: 546–554.

21. Abdel-Monem, H. (1978). *Essentials of Drug Product Quality*, C. V. Mosby, St. Louis, Mo., pp. 65–81.

22. Davidson, D. (1955). Amphoteric molecules, ions and salts. *J. Chem. Educ. 32*: 550–559.

23. Martin, A. N., Swarbrick, J., and Cammarata, A. (1970). *Physical Pharmacy*, 2nd ed. Lea & Febiger, Philadelphia, p. 291.

24. Hansch, C., Maloney, P. P., Fujita, T., and Muir, R. M. (1962). Correlation of biological activity of phenoxyacetic acids with Hammett substituent constants and partition coefficients. *Nature (London) 194*: 178.

25. Hansch, C., and Fujita, T. (1964). Rho-sigma-pi analysis. A method for the correlation of biological activity and chemical structure. *J. Am. Chem. Soc. 86*: 1616.

26. Fujita, T., Iwasa, J., and Hansch, C. (1964). A new substituent constant, pi, derived from partition coefficients. *J. Am. Chem. Soc. 86*: 5175.

27. Iwasa, J., Fujita, T., and Hansch, C. (1965). Substituent constants for aliphatic functions obtained from partition coefficeints. *J. Med. Chem. 8*: 150.

28. Hansch, C. (1960). A quantitative approach to biochemical structure-activity relationship. *Acc. Chem. Res. 2*: 232.

29. Hansch, C. (1971). Quantitative structure-activity relationship in drug design, in E. J. Ariëns, Ed., *Drug Design*, Vol. 1. Academic Press, New York, Chap. 2.

30. Lien, E. J. (1969). The use of substituent constants and regression analysis in the study of structure-activity relationship. *Am. J. Pharm. Educ., 33*, 368.

31. Tute, M. S. (1971). Principles and practice of Hansch analysis: a guide to structure-activity correlation for the medicinal chemist, in N. J. Harper and A. B. Simmonds, Eds., *Advances in Drug Reserach*, Vol. 6. Academic Press, New York, pp. 1–77.

32. Fujita, T. (1972). The extrathermodynamic structure-activity correlations: background of the Hansch approach,

in R. F. Gould, Ed. *Advances in Chemistry Series.* American Chemical Society, Washington, D.C., Chap. 1.

33. Martin, Y. C. (1978). *Quantitative Drug Design.* Marcel Dekker, New York.

34. Purcell, W. P., Bass, G. E., and Clayton, J. M. (1973). *Strategy of Drug Design: A Guide to Biological Activity.* Wiley, New York.

35. Hansch, C., and Leo, A. (1979). *Substituent Constants for Correlation Analysis in Chemistry and Biology.* Wiley, New York.

36. Salame, M. (1961). Prediction of liquid permeation in polyethylene and related polymers, *Soc. Plast. Eng. Trans. 1:* 153–163.

37. Salame, M., and Pinsky, J. (1962). Permeability prediction. *Mod. Packag. 36:* 153–156.

38. Glave, W. R., and Hansch, C. (1972). Relationship between lipophilic character and anesthetic activity. *J. Pharm. Sci. 61:* 589.

39. Bird, A. E., and Marshall, A. C. (1967). Correlation of serum binding of penicillins with partition coefficients. *Biochem. Pharmacol. 16:* 2275.

40. Vandenbelt, J. M., Hansch, C., and Church, C. (1972). Binding of apolar moelcules by serum albumin. *J. Med. Chem. 15:* 787.

41. Hansch, C., and Lien, E. J. (1968). An analysis of the structure-activity relationship in the adrenergic blocking activity of the β-haloalalkylamines. *Biochem. Pharmacol. 17:* 715.

42. Selassie, C. D., Wang, P. H.,a nd Lien, E. J. (1980). Reevaluation of bulk parameters: molar refraction, molecular mass, molar volume and parachor. *Acta Pharm. Jugosl. 30:* 135.

43. Wang, P. H., and Lien, E. J. (1980). Effects of different buffer species on partition coefficients of drugs used in quantitative structure-activity relationships. *J. Pharm. Sci. 69:* 662.

44. Lien, E. J., and Wang, P. H. (1980). Lipophilicity,

molecular weight and drug action: Reexamination of parabolic and bilinear models. *J. Pharm. Sci. 69:* 648.

45. Lien, E. J., Guo, Z. R., Li, R. L., and Su, C. T. (1982). Use of dipole moment as a parameter in drug-receptor interaction and quantitative structure-activity relationship studies. *J. Pharm. Sci. 71:* 641.

46. Lien, E. J., Tong, G. L., Chou, J. T., and Lien, L. L. (1973). Structural requirements for central acting drugs, I. *J. Pharm. Sci. 62:* 246.

47. Lien, E. J., Liao, R. C. H., and Shinouda, H. G. (1970). Quantitative structure-activity relationship and dipole moments of anticonvulsants and CNS depressants. *J. Pharm. Sci. 68:* 463.

48. Hansch, C., Steward, A. R., Anderson, S. M., and Bentley, D. (1968). The parabolic dependence of drug action upon lipophilic character as revealed by a study of hypnotics. *J. Med. Chem. 11:* 1.

49. Franke, R., and Oehme, P. (1973). Aktuelle Probleme der Wirkstofforschung. 1. Mitteilung: Ermittlung quantitative Structur-Wirkungs-Beziehungenbei Biowirkstoffen: Theoretische Grundlagen, Durchfuhrung. *Pharmazie 28:* 489.

50. Kubinyi, H. (1977). Quantitative structure-activity relationship of the bilinear model, a new model for nonlinear dependence of biological activity on hydrophobic character. *J. Med. Chem. 20:* 625.

51. Hyde, R. M. (1975). Relationships between the biological and physicochemical properties of series of compounds. *J. Med. Chem. 18:* 231.

52. Higuchi, T., and Davis, S. S. (1970). Thermodynamic analysis of structure-activity relationships of drugs: prediction of optimal structure. *J. Pharm. Sci. 59:* 1376.

53. Ho, N. F. H., Park, J. Y., Morozowich, W., and Higuchi, W. I. (1977). In, E. B. Roche, Ed., *Design of Biopharmaceutical Properties Through Prodrugs and Analogs.* APha Academy of Pharmaceutical Sciences, Washington, D.C., pp. 136–727.

54. Lien, E. J., Alhaider, A. A., and Lee, V. H. L. (1982). Phase partition: its use in the prediction of membrane permeation and drug action in the eye. *J. Parent. Sci. Technol. 36*: 86.

55. Li, W. Y., Guo, Z. R., and Lien, W. J. (1984). Examination of interrelationship between aliphatic group dipole moment and polar substituent constants. *J. Pharm. Sci. 73*: 553.

56. Lien, E. J. (1980). In, A. Osol, Ed., *Remington's Pharmaceutical Sciences*, 16th ed. Mack Publishing, Easton, Pa., pp. 160–181.

57. Lien, E. J. (1982). Molecular weight and therapeutic dose of drug. *J. Clin. Hosp. Pharm. 7*: 101–106.

58. Skolnik, P., and Paul, S. M. (1981). Benzodiazepine receptors, in H.-J. Hess, Ed., *Annual Reprots in Medicinal Chemistry*, Vol. 16, Academic Press, New York, pp. 21–26.

59. Lim, R. K. S., Guzman, F., Rodgers, D. W., Goto, K., Braun, C., Dickerson, C. D., and Engle, R. J. (1964). Site of action of narcotic and non-narcotic analgesics determined by bradykinin-evoked visceral pain. *Arch. Int. Pharmacodyn. Ther. 152*: 25–58.

60. Korolkovas, A., and Burckhalter, J. H. (1976). *Essentials of Medicinal Chemistry*. Wiley, New York, pp. 113–129.

61. Buckett, W. R. (1964). The relationship between analgesic activity, acute toxicity and chemical structure in esters of 14-hydroxycodeinone. *J. Pharm. Pharmacol. 16*, Suppl. 68T-71T.

62. Lien, E. J., Tong, G. L., Srulevitch, D. B., and Dias, C. (1978). QSAR of narcotic analgetic agents, in G. Barnett, M. Trsic, and R. Willette, Eds., *QuaSAR Research Monograph 22*. National Institute of Drug Abuse, Rockville, Md., pp. 186–196.

63. Pert, C. B., Snyder, S. H., and Portoghese, P. S. (1976). Correlation of opiate receptor affinity with analgetic effects of meperidine homologues. *J. Med. Chem. 19*: 1248–1250.

64. Oh-ishi, T., and May, E. L. (1973). *N*-Alkylnorketobemi-

dones with strong agonist weak antagonist properties. *J. Med. Chem. 16*: 1376–1378.

65. Wilson, R. S., Rogers, M. C., Pert, C. B., and Snyder, S. H. (1975). Homogogues of *N*-alkylnormeperidines. Correlation of receptor binding with analgesic potency. *J. Med. Chem. 18*: 240–242.

66. Larson, D. L., and Portoghese, P. S. (1976). Relationship between ED_{50} does and time-course brain levels of *N*-alkylnormeperidine homologues. *J. Med. Chem. 19*: 16–19.

67. Casy, A. F. (1982). In *Medicinal Research Reviews*, Vol. 2. Wiley, New York, pp. 167–192.

68. Yang, T. T., Srulevitch, D. B., and Lien, E. J. (1981). QSAR of synthetic narcotic analgetic agents related to fentanyl. *Acta Pharm. Jugosl. 31*: 171–182.

69. Rekker, R. F. (1977). *The Hydrophobic Fragment Constant.* Elsevier, New York.

70. Soudijn, W., and Wijngaarden, I. van (1977). Peptides: structure-activity relationships, a missing link between bioactive peptides and general synthetics, in J. A. Keverling Buisman, Ed., *Biological Activity and Chemical Structure.* Elsevier, Amsterdam, pp. 195–209.

71. Davis, J. (1984). *Endorphins: New Waves in Brain Chemistry.* Dial Press/Doubleday, New York.

72. Bondi, A. (1964). Van der Waals volume and radii. *J. Phys. Chem. 68*: 441.

73. Lien, E. J., Ariëns, E. J., and Beld, A. J. (1976). Quantitative correlations between chemical structure and affinity for acetylcholine receptors. *Eur. J. Pharmacol. 35*: 245–252.

74. Lien, E. J., and Kennon, L. (1985). Molecular structure, properties and states of matter, in *Remington's Pharmaceutical Sciences*, 17th ed. Mack Publishing, Easton, Pa., Chap. 13.

75. Creutzfeldt, W. Ed. (1977). Cimetidine, in *Proceedings of an International Symposium on Histamine H_2-Receptor Antagonists*, November 1977, Göttingen, Federal Republic of Germany, Excerpta Medica, Amsterdam.

76. Barnhart, E. R., Publisher (1985). *Physicians' Desk Reference.* Medical Economics, Oradell, N.J.

77. Ganellin, C. R. (1978). Chemistry and structure-activity relationships of H_2-receptor antagonists, in M. R. e Silva, Ed., *Handbook of Experimental Pharmacology,* Vol. XVIII/2. Springer-Verlag, New York, pp. 250—294.

78. Pfeiffer, C. C. (1956). Optical isomerism and pharmacological action, a generalization. *Science 124*: 29—31.

79. Ariëns, E. J. (1971). In E. Ariëns, Ed., *Drug Action,* Vol. 1. Academic Press, New York, pp. 149—193.

80. Lien, E. J., Rodrigues de Miranda, J. F., and Ariëns, E. J. (1976). Quantitative structure-activity correlation of optical isomers: a molecular basis for the Pfeiffer's rule. *Mol. Pharmacol. 12*: 598—604.

81. Jönsson, A. (1961). In W. Ruhland, Ed., *Handbuch der Pflanzenphysiologie,* Vol. 14. Springer, Heidelberg, pp. 959—1001.

82. Åberg, B. (1961). In *4th International Conference on Plant Growth Regulation,* Ames, Iowa, State University Press, Yonkers, N.Y., pp. 219—232.

83. Ariëns, E. J., Soudijn, W., and Timmermans, P. B. M. W. M., Eds. (1983). *Stereochemistry and Biological Activity of Drugs.* Blackwell Scientific, Oxford.

84. Lien, E. J., Beld, A. J., and Ariëns, E. J. Quantitative structure-activity correlation of conformationally restricted antihistamines to histamine receptors, unpublished data.

85. Ison, R. R., Franks, F. M., and Soh, K. S. (1973). The binding of conformationally restricted antihistamines to histamine receptors. *J. Pharm. Pharmacol. Ther. 25*: 887—894.

86. Ariëns, E. J., and de Groot, W. M. (1954). Affinity and intrinsic-activity in the theory of competitive inhibition. Part III. Homologous decamethonium-derivatives and succinyl-choline-esters. *Arch. Int. Pharmacodyn. Ther. 99*: 193—205.

87. Van Rossum, J. M., and Van den Brink, F. G. (1963). Cumulative dose-response curves. I. Introduction to the technique. *Arch. Int. Pharmacodyn. Ther. 143*: 240—246.

88. Van Rossum, J. M. (1963). Cumulative dose-response curves. II. Technique for the making of dose-response curves in isolated organs and the evaluation of drug parameters. *Arch. Int. Pharmacodyn. 143*: 299–330.

89. Lehman, F. P. A., Rodrigeus de Miranda, J. F., and Ariëns, E. J. (1976). Stereoselectivity and affinity in molecular pharmacology. 1. The correlation of stereoselectivity and activity, a survey. In E. M. Jucker, ed., *Progress in Drug Research*, Vol. 20, pp. 101–126.

90. Rodrigues de Miranda, J. F., Lehman, P. A., and Ariëns, E. J. (1976). A molecular basis for eudismic-affinity correlations, part 2. In E. M. Jucker, ed., *Progress in Drug Research*, Vol. 20, pp. 126–142.

91. Ison, R. R., and Casy, A. F. (1971). Structural influence upon antihistamine activity; 3-amino-1-aryl-1-(2-pyridyl)-propenes and related compounds. *J. Pharm. Pharmacol. 23*: 848.

92. Laarhoven, W. H., Nivard, R. J. F., and Havinga, E. (1961). Estrogenic activity and steric hindrance to co-polanarity of alkylsubstituted 4,4'-dimethoxystilbenes. *Experientia 17*: 214–215.

93. Lien, E. J., Kuwahara, J., and Koda, R. T. (1974). Drug distribution and physicochemical properties: Diffusion of drugs into prostatic fluid and milk. *Drug Intell. Clin. Pharm. 8*: 470–475.

94. Szász, G. (1985). *Pharmaceutical Chemistry of Adrenergic and Cholinergic Drugs*. CRC Press, Boca Raton, Fla.

95. Liu, H. K., and Chiou, G. C. Y. (1981). Continuous, simultaneous, and instant display of aqueous humor dynamics with a micro spectrophotometer and a sensitive drop counter. *Exp. Eye Res. 32*: 583–592.

96. Keates, E. U., and Stone, R. (1984). The effect of d-timolol on intraocular pressure in patients with ocular hypertension. *Am. J. Ophthalmol. 98*: 73–78.

Four

Ribonucleotide Reductase Inhibitors as Antiviral and Anticancer Agents

I. RIBONUCLEOTIDE REDUCTASE (RIBONUCLEOSIDE DIPHOSPHATE REDUCTASE) AS A BIOCHEMICAL TARGET

Ribonucleotide reductase (RR) is a key enzyme involved in conversion of ribonucleotides to deoxyribonucleotides, a rate-limiting step in DNA synthesis and cell replication. This enzyme is, therefore, a potential target for the development of anticancer agents (1) (Figure 20).

There is a close correlation between the activity of this enzyme and cell replication (2, 3). The activity is low in slow-growing cells and high is rapidly growing and malignant cells

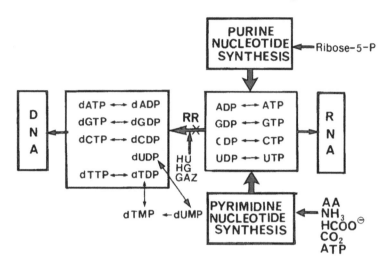

Figure 20 Role of ribonucleotide reductase (RR) in the synthesis of DNA. (Adapted from Ref. 1).

TABLE 32 Comparison of the Differential Activities of Some Enzymes Important in DNA Synthesis in Tumor Cells Compared to Normal Cells

Enzyme	Activity in hepatoma 3683-F (% of normal liver)
Ribonucleotide reductase	20,800
Thymidine kinase	10,200
DNA polymerase	5,800
Thymidine monophosphate synthetase	2,860
Dihydroorotase	418
Orotate phosphoribosyl transferase	328
Nucleoside diphosphate kinase	200

Source: Adapted from Ref. 7.

(e.g., sarcomas, hepatomas, thymus, spleen, and bone marrow cells) (4). Cells induced to synthesize DNA by virus infection also increase their ribonucleotide reductase activity rapidly (5, 6) (Table 32).

The pool size of deoxyribonucleotides is small, therefore cannot support DNA synthesis for any length of time; they must be formed just before incorporation into DNA (8). These findings point to ribonucleotide reductase as a good biochemical target for developing agents whose ultimate goal is the inhibition of de novo DNA synthesis.

In the past 10 years we have synthesized and tested some potential anticancer and antiviral agents whose biochemical target is the enzyme ribonucleotide reductase. It is worth noting that anticancer agents interfering with preformed DNA (e.g., alkylating agents, nitrosoureas, procarbazine, adriamycin, etc.) are also well-known carcinogens, while drugs such as hydroxyurea and methotrexate which inhibit the de novo synthesis of DNA are not carcinogenic under normal therapeutic conditions (see Figure 21).

Among the different ribonucleotide reductases from different sources, the enzyme from *Escherichia coli* has been most extensively studied and well characterized. Table 27 summarizes the various characteristics and cofactors required for the RRs from different sources.

Figure 22A is a current model of *E. coli* ribonucleotide reductase proposed by Reichard and Ehrenberg (18) showing the presence of the —SH groups on the B_1 subunit and the tyrosine free radical and Fe^{3+} on the B_2 subunit. Figure 22B is a model of *Lactobacillus leichmanii* enzyme proposed by Thelander (18), showing the regulator site away from the catalytic site.

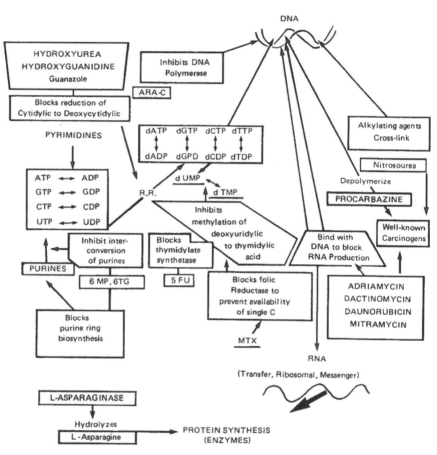

Figure 21 Mechanisms of antitumor agents interfering with cell replication. ·

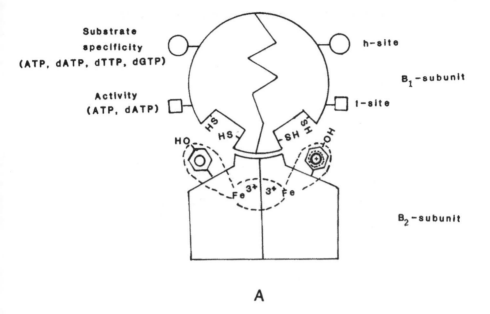

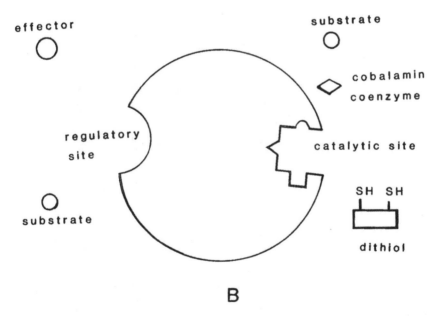

Figure 22 (A) Model of *E. coli* ribonucleotide reductase compared to (B) model of *L. leichmanii* ribonucleotide reductase.

TABLE 33 Comparison of Ribonucleotide Reductases from Different Sources

Source	E. coli	L. leichmanii	Rat novikiff hepatoma	Calf thymus	Ehrlich ascites
Substrate	NDP	NTP	NDP	NDP	NDP
Cofactors	Fe, ATP, Mg	Coenzyme B_{12}	Fe, ATP, Mg	Fe	
Purification	Homogeneous	Homogeneous	150X	3400X	
MW	B1 160,000 B2 78,000	76,000	?	M1 84,000 M2 11,000	
Subunits	B1 and B2	Monomeric	P1 and P2	M1 and M2	Tris fraction and dye fraction (similar to B1 and B2 from E. coli)

Requirement for Mg	Absolute	Relative	Absolute	Absolute	Absolute
Requirement for B$_{12}$	No	Yes	No	No	No
Involvement of Fe	Yes	No	Yes	Yes	Yes
Physiologic H donor	Thioredoxin	Thioredoxin	Thioredoxin	Thioredoxin	Thioredoxin
Activator	ATP	ATP	ATP	ATP	ATP
Inhibitor	dATP	None	dATP	dATP	dATP
Hydrogen transport system	TPNH → FADH → (—SH)$_2$	TPNH → FADH → (—SH)$_2$	TPHN → enzymes	TPNH → FADH → (—SH)$_2$	TPNH → FADH → (—SH)$_2$
References	10–12	13, 14	15	16	17

The coupling of RR to thioredoxin and the oxidation reduction of the active disulfide in the *E. coli* enzyme is shown in Figure 23. The hydrogen donor system in the *E. coli* reductase and the lactobacillus reductase are compared in Figure 24 (18).

II. INHIBITORS OF RR

Hydroxyurea, hydroxyguanidine, the thiosemicarbazones, and guanazole have all been shown to have anticancer activities and in some cases antiviral activities. There is evidence that the site of action is at the level of the enzyme ribonucleotide reductase. Although these compounds are all active, their clinical use is limited either by an inability to reach the target site in vivo or due to toxic effects. Hydroxyurea is used in the treatment of leukemias and head and neck cancers. It is the only drug used clinically whose primary mode of action is the inhibition of ribonucleotide reductase. The enzyme contains Fe^{2+} as a cofactor, and one of the possible mechanisms of action of inhibitors such as hydroxyurea is the ability to abolish a free radical or to form a chelate.

Guanidine is an antiviral agent. The viral etiology of some cancers have been shown by the presence of reverse transcriptase in human cells (19). Hydroxyguanidine was designed on the basis that it combines the essential moieties of molecules with antitumor activity (the hydroxyamino group of hydroxyurea) and those with antiviral activity (the imino group of guanidine) in the hope of obtaining a compound with both types of activities (20).

The α-(N)-heterocyclic thiosemicarbazones have been studies by several groups (21–26) and found to possess both antineoplastic and antiviral activities. Brockman (23, 24) first

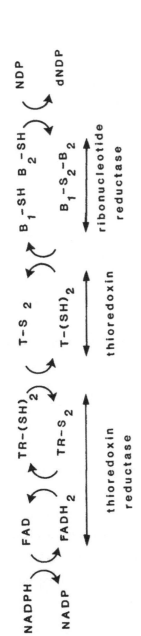

Figure 23 Oxidation reduction of active disulfides in ribonucleotide reduction in *E. coli*.

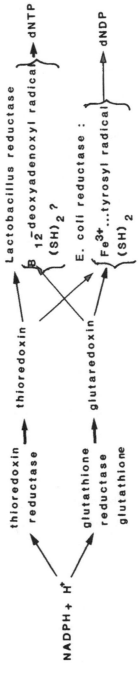

Figure 24 Hydrogen donor systems in ribonucleotide reduction.

discovered that 2-formylpyridine thiosemicarbazone increased the life span of mice with L1210 leukemia. French and Blanz (25) found that 1-formylisoquinoline thiosemicarbazone was active against L1210 and other tumors. Sartorelli (26) showed that 1 formylisoquinoline thiosemicarbazone inhibits DNA synthesis in mice with L1210 leukemia and also binds reversibly to ribonucleotide reductase from Novikoff rat tumor. Agrawal and Sartorelli (21) made extensive modifications of those compounds both on the ring structure and in the side chain. Many of the compounds synthesized were active in vitro but not in vivo or had a low therapeutic index. Methylisatin-3-thiosemicarbazone is active against a wide spectrum of viruses (e.g., herpes simplex virus, human cytomegalovirus, and vaccinia virus). It has been used for the treatment of vaccinia infections and in the prevention of smallpox (22). It acts by inhibiting DNA-dependent RNA synthesis, and its antiviral activity is closely correlated with its ability to inhibit ribonucleotide reductase.

The mechanism of action of this group of compounds is that either they chelate Fe^{2+} necessary for enzyme activity, or an iron complex of the compound blocks the enzyme or by a free radial mechanism.

These considerations led us to explore the synthesis of some 1-amino-3-hydroxyquanidine derivatives with structural features of both hydroxyguanidine and the thiosemicarbazones. The aim is to optimize both anticancer and antiviral activities. Our preliminary results have been encouraging. Based on these results and on QSAR analysis, we propose to synthesize some more compounds with different physicochemical properties which would not only give more information on the structural requirements of the active site but may lead to better chemotherapeutic agents.

Guanazole (3,5-diamino-1,2,4-triazole) has been shown to have antitumor activity against murine leukemia L1210, W256 carcinoma, some human lymphoblast cell lines, and other cancer cells. It has been established that guanazole also acts by inhibitng a ribonucleotide reductase and therefore DNA synthesis (23). Clinical trials of this drug by continuous introvenous (i.v.) infusion have achieved remissions in some patients with leukemia, but the short half-life in humans (1 to 2 hr) limits its use. Hahn et al. (28) performed some structure-activity relationship studies of the effects of compounds related to guanazole against L1210 leukemia in vitro. They found that elimination of both the 3- and 5-amino groups of guanazole abolished activity, elimination of only one amino group decreased but did not abolish activity, and substitution of a sulfhydryl group for an amino group in the 3-poisition (3-mercapto-1,2,3-triazole) produced a very active compound in vitro but not in vivo. Other substitutions on the ring structure or changes in the ring structure led to compounds with decreased activity, or with equal activity in vitro, but inactive or toxic in vivo. The limiting factor of guanazole is its hydrophilicity and low molecular weight; therefore, like hydroxyurea, it needs to be dosed frequently.

Hydroxyurea (HU) is a clinically used anticancer drug, especially in the treatment of chronic granulocytic leukemia as well as head and neck cancer (29–30). Young et al. (32) found that the essential pharmacophore in the hydroxyurea molecule is the hydroxamic acid moiety $-\text{N}-\text{C}\overset{\diagup}{\diagdown}$.

$$\underset{\text{OH}}{|}$$

TABLE 34 Biological Activity of HG Derivatives

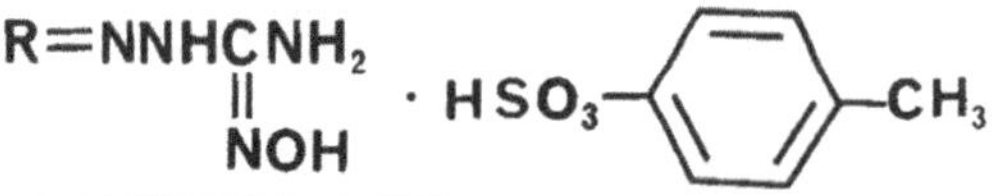

$$R=NNHCNH_2 \cdot HSO_3- \text{(benzene ring)} -CH_3$$
$$\quad\quad \underset{NOH}{\overset{\parallel}{}}$$

Compound R =	Inhibition of RR[a]: ID_{50} M	Inhibition of L1210 in vitro[b]: $ID_{50} \times 10^{-6}$ M	Inhibition of RSV[b]: $ID_{50} \times 10^{-6}$ M
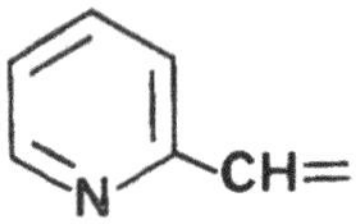 (pyridin-2-yl)CH=	3.8×10^{-5}	10.91	2.76
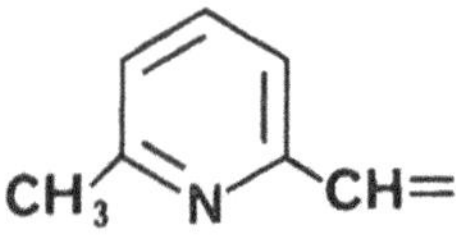 (6-methylpyridin-2-yl)CH=		42.08	114.10
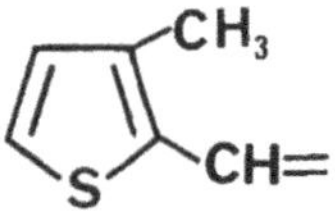 (3-methylthiophen-2-yl)CH=	7.0×10^{-5}	33.22	18.46
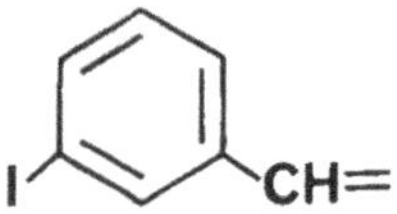 (3-iodophenyl)CH=	3.1×10^{-5}	12.52	3.45
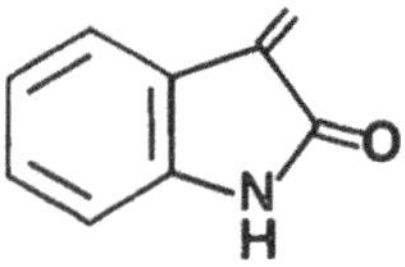 (oxindolylidene)		48.39	18.27
$H_2=$		126.00	195.20
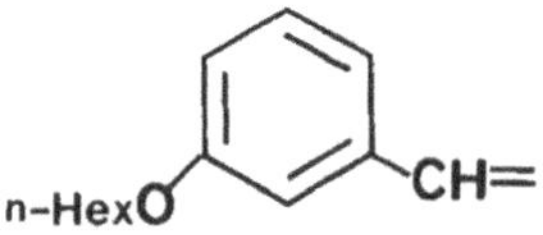 n-HexO-phenyl-CH=		7.80	8.02
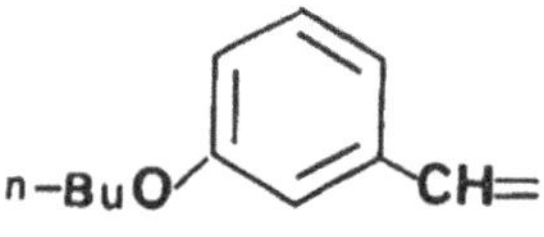 n-BuO-phenyl-CH=		8.54	10.18

TABLE 34 (Continued).

Compound R =	Inhibition of RR[a]: ID_{50} M	Inhibition of L1210 in vitro[b]: $ID_{50} \times 10^{-6}$ M	Inhibition of RSV[b]: $ID_{50} \times 10^{-6}$ M
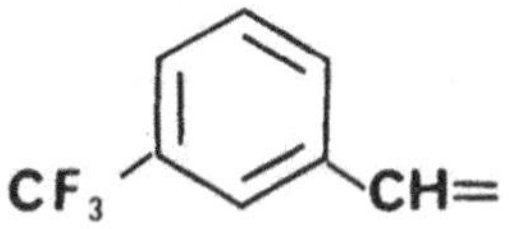 CF_3 — benzene — CH=		15.1	18.24
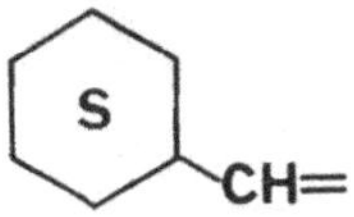 S — ring — CH=		21.24	23.45
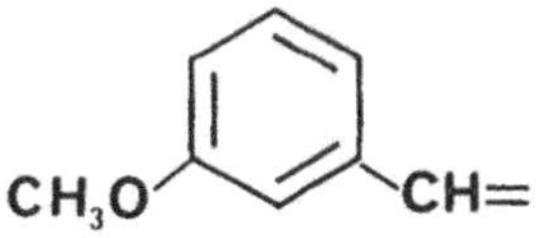 CH_3O — benzene — CH=		16.00	17.75
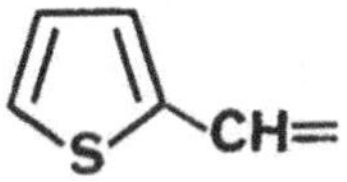 thiophene — CH=		34.11	11.77
Hydroxyxguanidine sulfate	9×10^{-4}	59.49	374.00
Hydroxyurea		60.02	

[a]From Ref. 33.
[b]From Ref. 34.

The anticancer activity of hydroxyurea is generally based on its ability to inhibit the enzyme involved in DNA synthesis, ribonucleotide reductase. This enzyme catalyzes the reduction of ribonucleotides to deoxyribonucleotides which eventually incorporate in the DNA molecule.

Our preliminary studies with 11 hydroxyguanidine derivatives are very promising (33, 34) (Table 34). The ID_{50} of
these compounds against L1210 leukemia cells in vitro were in
the range 7.8 to 126 μM. The ID_{50} of six of these compounds
tested against CCRF-CEM human leukemia cells were in the range
2.8 to 36 μM (9). All the compounds were found to be more
active than both hydroxyguanidine and hydroxyurea. The most
active compounds were about 10 times more active. The ID_{50}
against Rous sarcome virus were in the range 2.76 to 195.2 μM,
compared to 374 μM for hydroxyguanidine. The most active
compound was 100-fold more active than hydroxyguanidine in
in vitro antiviral tests.

The preliminary study with the N—OH imide derivatives
shows that the best compound in the small series 4-amino-N-
benzenesulfonyloxyphthalmide has an ID_{50} of 2.28×10^{-6} M,
while hydroxyurea has an ID_{50} of about 10^{-3} M (Table 35).

The character of the tautomeric equilibria and the pK_a
values of some selected N-amino-N'-hydroxyguanidine derivatives have been determined by ^{15}N NMR spectroscopy in cal-
laboration with J. D. Roberts of the California Institute of
Technology (Table 36).

QSAR analysis of the N-amino-N'-hydroxyguanidine derivatives (34) compounds gave the following equations:

Inhibition of L1210 leukemia cells:

$$\log \frac{1}{ID_{50}} = 0.752 \log MR + 3.587$$

n	r	s
11	0.829	0.212

Inhibition of Rous sarcoma virus:

$$\log \frac{1}{ID_{50}} = 1.181 \log MW_R + 1.896R_m$$
$$+ 22.71$$

	n	r	s
	11	0.919	0.215

TABLE 35 Anticancer Activity of N—OH Imide Derivatives

Compound	Inhibition of L1210/0[a]
	ID_{50} (48 hr) $\times$ 10^{-6} M

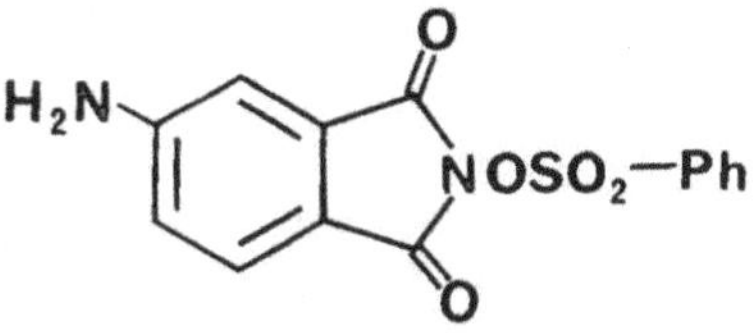

Compound	Inhibition of L1210/0[a]
H₂N-[structure]-NOSO₂—Ph	2.28
[structure]-NOSO₂—Ph	4.42
[structure]-N—OH	11.7
H₂N-[structure]-N—OH	179
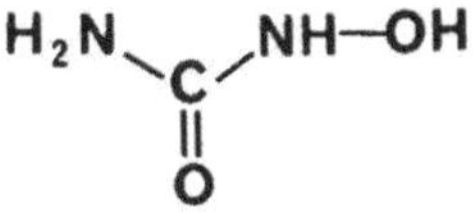 H₂N-C(=O)-NH—OH (HU)	>100 (ca. 1000)

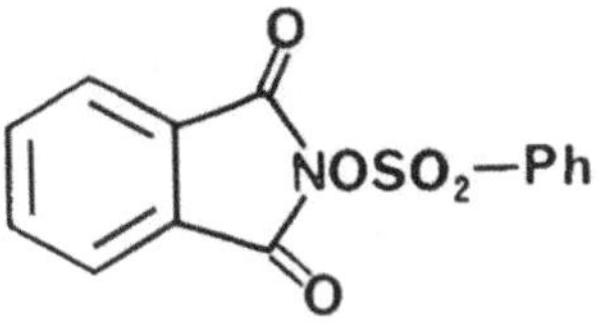

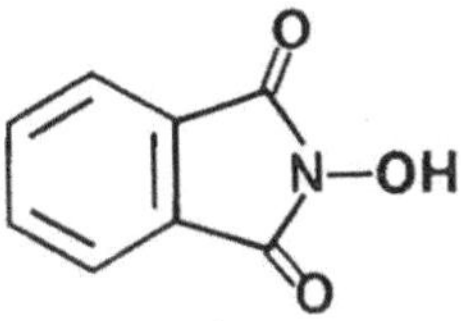

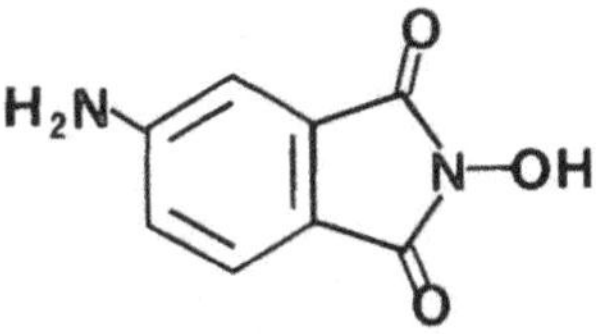

[a]From Ref. 35.

The correlation with log MW and log MR indicates that increased bulk is important. The positive dependence on R_m indicates that increased hydrophobicity is advantageous. We could not obtain good correlation with electronic properties.

TABLE 36 ^{15}N Chemical Shifts for Salts of Aminohydroxyguanidine Derivatives in $(CH_3)_2SO$[a]

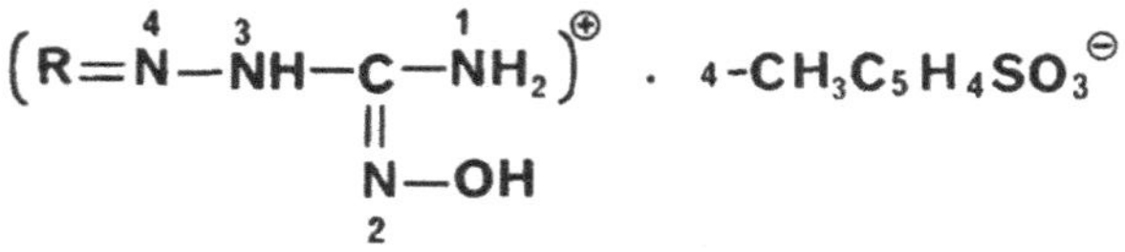

$$\left(R{=}\overset{4}{N}{-}\overset{3}{NH}{-}\overset{1}{C}{-}NH_2\right)^{\oplus} \cdot \ 4\text{-}CH_3C_5H_4SO_3^{\ominus}$$
$$\underset{2}{\overset{\|}{N}}{-}OH$$

Compound	N1	N2	N3	N4	R =	pK_a
1	300.6	235.0	237.9	58.0	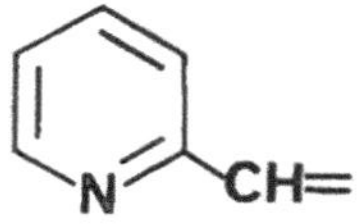 pyridin-2-yl—CH=	6.7 ± 0.1
2	302.0	235.9	239.2	57.4	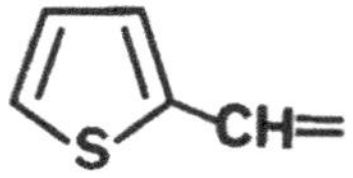 thiophen-2-yl—CH=	—
3	300.6	235.5	238.3	—[b]	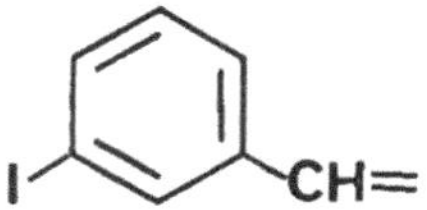 3-iodophenyl—CH=	—
4	305.9	241.6	280.5	320.9	H_2=	8.7 ± 0.1

[a]In ppm upfield from 1 M $H^{15}NO_3$ in D_2O as external standard.
[b]Not observed.
Source: Ref. 33.

This could be due to the small range in electronic properties in this series of compounds. Dunn et al. (36) found that electron withdrawal decreased the activity of the α-(N)-heterocyclic carboxaldehyde thiosemicarbazones and attributed this to destabilization of chelate formation.

Various biotargets, including ribonucleotide reductase, for the design of antiviral agents have been compiled by De Clercq and Walker (37).

REFERENCES

1. Sartorelli, A. C., and Agrawal, K. C. (1976). In A. C. Sartorelli, Ed., *Cancer Chemotherapy*, ACS Symposium Series 30. American Chemical Society, Washington, D. C., pp. 1–14.

2. Cory, J. G., and Whitford, T. W., Jr. (1972). Ribonucleotide reductase and DNA synthesis in Ehrlich ascites tumor cells. *Cancer Res. 32*: 1301–1306.

3. Elford, H. L., Wampler, G. L., and van't Riet, B. (1979). New ribonucleotide reductase inhibitors with antineoplastic activity. *Cancer Res. 39*: 844–851.

4. Takeda, E., and Weber, G. (1981). Role of ribonucleotide reductase in expression of the neoplastic program. *Life Sci. 28*: 1007–1014.

5. Langelier, Y., Dechamps, M., and Buttin, G. (1978). Analysis of dCMP deaminase and CDP reductase levels in Hamster cells infected by herpes simplex virus. *J. Virol. 26*: 547–553.

6. Lindberg, U., Nordenskjöld, B. A., Reichard, P., and Skoog, L. (1978). Thymidine phosphate pools and DNA synthesis after polyoma infection of mouse embryo cells. *Cancer Res. 29*: 1498–1506.

7. Weber, G. (1977) Enzymology of cancer cells. *N. Engl. J. Med. 296*: 486–493.

8. Walters, R. A., Tobey, R. A., and Ratliff, R. L. (1973).

Cell cycle dependent variations of deoxyribonucleoside triphosphate pools in Chinese hamster cells. *Biochim. Biophys. Acta.* *319*: 336–347.

9. Tai, A. W. (1982). Design of novel 1-amino-3-hydroxy-guanidine derivatives as antiviral and anticancer agents, Ph.D. dissertation, University of Southern California, School of Pharmacy, Los Angeles.

10. Larsson, A., and Reichard, P. (1966). Enzymatic synthesis of deoxyribonucleotides. *J. Biol. Chem.* *241*: 2533–2539.

11. Larsson, A., and Reichard, P. (1966). Enzymatic synthesis of deoxyribonucleotides. X. Reduction of purine ribonucleotides, allosteric behavior and substrate specificity of the enzyme system. *J. Biol. Chem.* *241*: 2540–2549.

12. Laurent, T. C., Moore, E. C., and Reichard, P. (1964). Enzymatic synthesis of deoxyribonucleotides. Isolation and characterization of thioredoxin, the hydrogen donor from *Escherichia coli B*. *J. Biol. Chem.* *239*: 2436–2444.

13. Goulian, M., and Beck, W. S. (1966). Purification and properties of cobamide-dependent ribonucleotide reductase from *Lactobacillus leichmannii*. *J. Biol. Chem.* *241*: 4233–4242.

14. Beck, W. S., Goulian, M., Larsson, A., and Reichard, P. (1966). Hydrogen donor specificity of cobamide dependent ribonucleotide reductase and allosteric regulation of substrate specificity. *J. Biol. Chem.* *241*: 2177–2179.

15. Moore, E. C. (1977). Components and control of ribonucleotide reductase system of the rat. *Adv. Enzyme Regul.* *15*: 101–114.

16. Engstrom, Y., Eriksson, S., Thelander, L., and Akerman, M. (1979). Ribonucleotide reductase from calf thymus. Purification and properties. *Biochemistry* *18*: 2841–2948.

17. Cory, J. G. (1979). Properties of ribonucleotide reductase from Ehrlich tumor cells, multiple nucleoside diphosphate activities and reconstitution of activity from components. *Adv. Enzyme Regul.* *17*: 115–131.

18. T'ang, A. (1984). Optimization of ribonucleotide reductase inhibitors as anticancer and antiviral agents: *N*-hydroxy-*N'*-aminoguanidine derivatives, Ph.D. dissertation, Univer-

sity of Southern California, School of Pharmacy, Los
Angeles.

19. Gallo, R. C., Yang, S. S., and Ting, R. C. (1970).
RNA-dependent DNA polymerase of human acute leukemic
cells. *Nature (London)* 228: 927—929.

20. Adamson, R. H. (1972). Hydroxyguanidine—a new antitumor drug. *Nature (London)* 236: 400—401.

21. Agrawal, K. C., and Sartorelli, A. C. (1978). Chemistry
and biological activity of (α)-heterocyclic carboxaldehyde
thiosemicarbazone. *Prog. Med. Chem.* 15: 321—356.

22. Bauer, D. J., Ed. (1972). *Chemotherapy of Virus Diseases*, Vol. 1. Pergamon Press, Elmsford, N.Y.

23. Brockman, R. W., Shaddix, S., Laster, W. R., Jr.,
and Schabel, F. M., Jr. (1970). Inhibition of ribonucleotide reductase, DNA synthesis, and L1210 leukemia by
guanuzale. *Cancer Res.* 30: 2358—2368.

24. Brockman, R. W. Thomson, J. R., Bell, M. J., and
Skipper, H. E. (1956). Observations on the antileukemic
activity of pyridine-2-carboxaldehyde thiosemicarbazone
and thiocarbohydrazone. *Cancer Res.* 16: 167—170.

25. French, F. A., and Blanz, E. J., Jr. (1965). The car-
cinostatic activity of α-(N) heterocyclic carboxaldehyde
thiosemicarbazones. I. Isoquinoline-1-carboxaldehyde
thiosemicarbazone. *Cancer Res.* 25: 1454—1458.

26. Sartorelli, A. C., Agrawal, K. C., and Moore, E. C.
(1971). Mechanism of inhibition of ribonucleoside diphos-
phates by α-(N)-heterocyclic aldehyde tiosemicarbazones.
Biochem. Pharmacol. 20: 3119—3123.

27. Sartorelli, A. C. (1967). Effect of chelating agents,
upon the synthesis of nucleic acids and protein: inhibition
of DNA synthesis by 1-formylisoquinoline thiosemicarbazone.
Biochem. Biophys. Res. Commun. 27: 26—32.

28. Hahn, M. A., and Adamson, R. H. (1972). Pharmacology
of 3,5-diamino-1,2,4-traizole (guanazole). 1. Antitumor
activity of guanazole. *J. Nat. Cancer Inst.* 48: 783—790.

29. Kennedy, B. J., and Yarbo, J. W. (1966). Metabolic and
therapeutic effects of hydroxyurea in chronic myeloid
leukemia. *J. Am. Med. Assoc.* 195: 1038—1043.

30. Beckloff, G. L., and Lerner, H. J. (1967). Concomitant use of hydroxyurea and x-irradiation in treatment of head and neck cancer. *Int. Cong. Chemother.* 5: 353–359.

31. Watne, A., and Furner, P. (1968). Combination drug and x-ray therapy. *Proc. Am. Assoc. Cancer Res.* 9: 75.

32. Young, C. W., Schochetman, G., and Karnofsky, C. (1967). Hydroxyurea-induced inhibtion of deoxyribonucleotide synthesis: studies in intact cell. *Cancer Res.* 27: 526–534.

33. Tai, A. W., Lien, E. J., Moore, E. C., Chun, Y., and Roberts, J. D. (1983). Studies of N-hydroxy-N'-aminoguanidine derivatives by nitrogen-15 nuclear magnetic resonance spectroscopy and as ribonucleotide reductase inhibitors. *J. Med. Chem.* 26: 1326–1329.

34. Tai, A. W., Lien, E. J., Lai, M. M. C., and Khwaja, T. A. (1984). Novel N-hydroxyguanidine derivatives as anticancer and antiviral agents. *J. Med. Chem.* 27: 236–238.

35. Selassie, D., and Lien, E. J. (1985). Substituted-N-benzenesulfonyloxyphthalimide, unpublished data.

36. Dunn, W. J., and Hodnett, E. M. (1977). Structure-activity relationships for the inhibtion of ribonucleoside diphosphate reductase by α-(N)-formylheteroaromatic thiosemicarbazones. *Eur. J. Med. Chem.* 12: 113–116.

37. De Clercq, E., and Walker, R. T. Eds. (1982). *Targets for the Design of Antiviral Agents.* Plenum Press, New York.

Five

Chemical Structure and Side Effects

Any effect of a drug, desired or undesired, direct or indirect,
is determined by its physicochemical properties, which, in turn,
are governed by its chemical structure.   Other intrinsic factors,
such as genetic factors, sex, age, physiological and pathological
state of the patient, and extrinsic factors, such as environmental

conditions and time and frequency of exposure to the drug, are also extremely important, but are considered beyond the scope of this work (Figure 25). The data were taken from the literature with additional information added to the previous reports (1–11).

There are several dozen antihistaminics available commercially. It is not easy for clinicians to select the best drug for particular patients from these many products. In addition to the primary consideration of effectiveness, the incidence of side effects should be considered in choosing the best drug for each patient. CNS depression is the most common untoward effect observed for antihistaminics. Among the most potent and highly sedative of these drugs are the ethanolamine ($-O-CH_2CH_2N=$) derivatives [e.g., diphenhydramine (Benadryl), doxylamine (Decapryn)] and the phenothiazine derivatives [e.g., prometh-

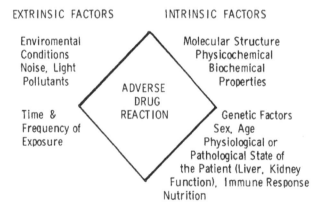

EXTRINSIC FACTORS   INTRINSIC FACTORS

Enviromental   Molecular Structure
Conditions   Physicochemical
Noise, Light   Biochemical
Pollutants   Properties
            ADVERSE
            DRUG
Time &   REACTION   Genetic Factors
Frequency of   Sex, Age
Exposure   Physiological or
                             Pathological State of
                             the Patient (Liver, Kidney
                             Function), Immune Response
                             Nutrition

Figure 25   Various intrinsic and extrinsic factors affecting adverse drug reaction.

azine (Phenergan)]. However, other drugs of these subgroups [e.g., carbinoxamine (Clistin), methdialazine (Tacaryl)] produce a relatively low incidence of CNS depression (12).

Ethylenediamine ($=NCH_2CH_2H=$) derivatives such as tripelenamine (Pyribenzamine) and chlorothen (Tagathen) also produce sedation. Alkylamines, such as chlorpheniramine (Chlor-Trimeton), in general have the lowest sedative side effects (12) (see Table 37). The CNS effects will be very minimal especially if one uses an active isomer such as dexchlorpheniramine maleate (Polaramine), since only 2 mg is required to achieve the H_1 blockade effect.

Terfenadine (Seldane), a piperidinebutanol derivative, is chemically equivalent to two butanol moieties attached to one nitrogen, with one of the butanol functions incorporated into the piperidine ring. It is the first peripheral H_1-specific antagonist marketed in the United States, with practically no CNS sedation side effect (13—16) (30):

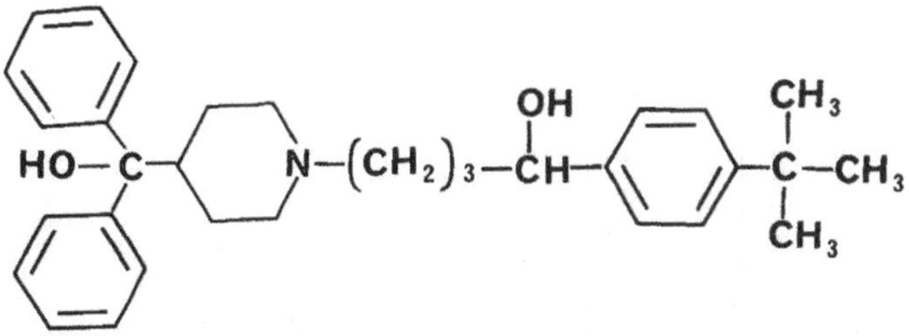

(30)

Terfenadine
α-[4-(1,1-Diemthylethyl)phenyl]-4-(hydroxydiphenylmethyl)-1-piperidinebutanol

TABLE 37 Comparison of H_1-Antihistamine Properties Among Different Groups

	H_1 blockade	Sedative effects (CNS)	Anticholinergic effects	Gastrointestinal effects
Ethylenediamines	+++	++	+	+++
Ethanolamines	+++	+++	+++	+
Alkylamines	+++	+	+	+
Piperazines	+++	++	+	+
Piperidines	+++	++	+	+
Phenothiazines	+++	+++	+++	+

Source: Adapted from Ref. 12.

Phenindamine (Nolahist), a tetrahydro-2-azafluorene derivative (with the propylamine group incorporated into the ring), is less likely to cause sedation and may even cause stimulation (31):

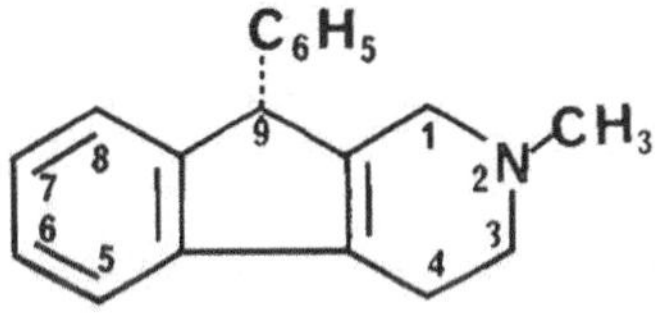

(31)

Phenindamine (2,3,4,9-tetrahydro-2-methyl-9-phenyl-
1H-indeno[2,1-c]pyridine) (1,2,3,4-tetrahydro-2-methyl-
9-phenyl-2-azafluorene)

Poperizine derivatives such as meclizine, cyclizine, and chlorcyclizine have been found to be teratogenic in experimental animals (2). Therefore, these drugs should not be prescribed for women of childbearing ages (32, 33).

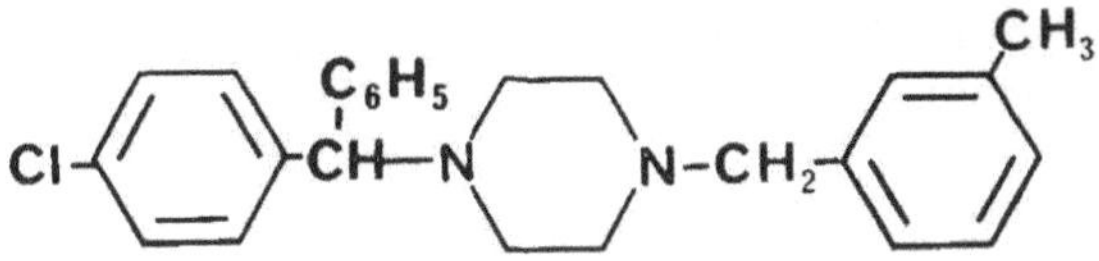

(32)

Meclizine

1-(p-Chloro-α-Phenylbenzyl)4-
(m-methylbenzyl)piperazine

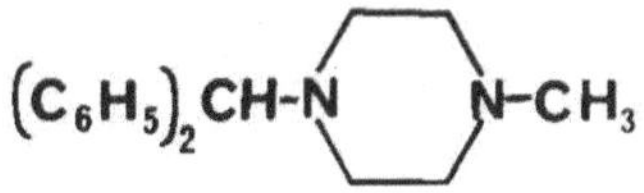

(33)

Cyclizine

1-Diphenyl-4-methylpiperazine

I. EXTRAPYRAMIDAL SYNDROME

Using a computer program to sort the data bank of several hundred clinically useful drugs, we (3) have found the following striking features common to the drugs producing extrapyramidal syndrome: a phenothiazine ring with a tertiary amino group separated from the ring by three carbon atoms, such as chlorpromazine, promazine, trifluoperazine, fluphenazine, mesoridazine, thiorodazine, thiopropazate, acetophenazine, carphenazine, and piperacetazine (34, 35, and Figure 26).

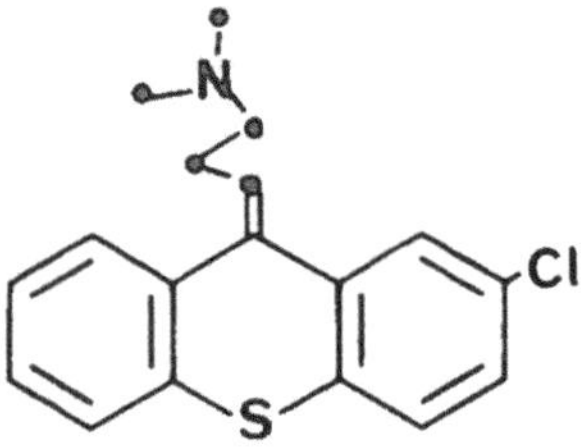

(34)

2-Chloro-10-[3-(dimethylamino)-
propyl]phenothiazine

Chlorpromazine

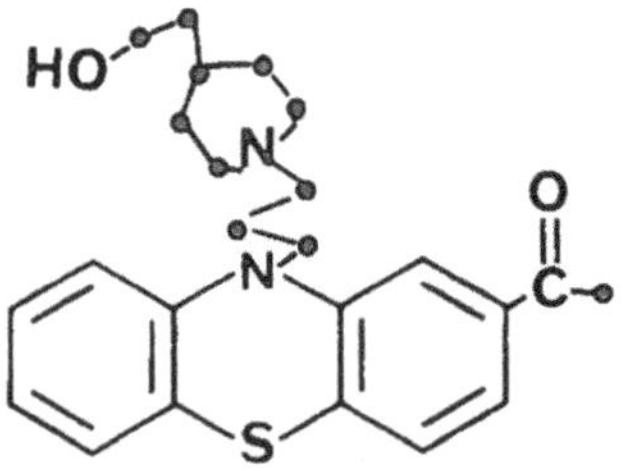

(35)

10-{3-[4-(2-hydroxyethyl)-1-piperazinyl]-
propyl}-2-propionylphenothiazine

Carphenazine

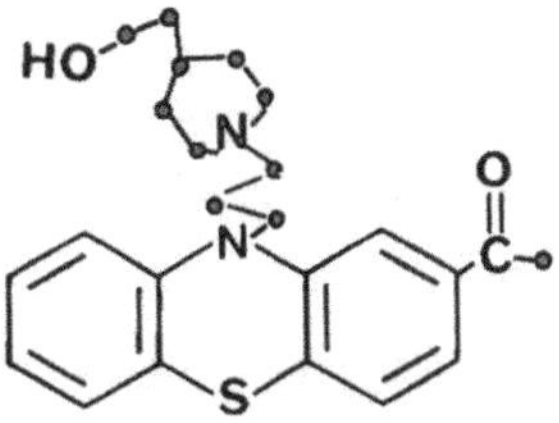

2-Acetyl-10-{3-[4-(β-hydroxyethyl)piperidino]propyl}-
phenothiazine

Piperacetazine

Figure 26 Structural formulas of typical phenothiazine derivatives with a tertiary amino group separated from the phenothiazine ring system by three carbon atoms. The hydrogens on
the carbon atoms are not shown.

Isosteres of phenothiazine, such as chlorprothixene, doxepin,

and thiothexine, are also known to cause the same type of side

effect (Figure 27).

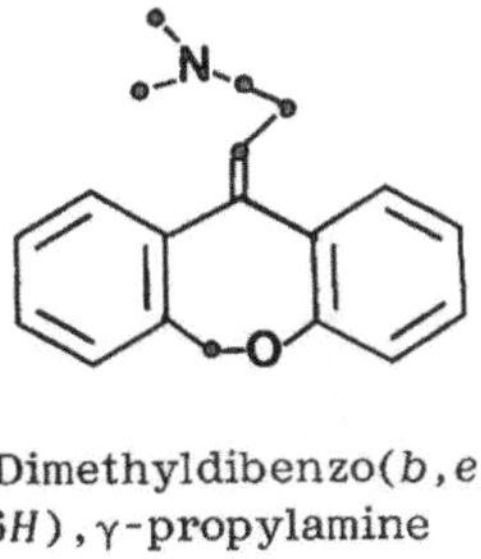

N,N-Dimethyldibenzo(b,e)oxepin-   2-Chloro-N,N-dimethylthioxanthene-
Δ11(6H),γ-propylamine   Δ9-γ-propylamine

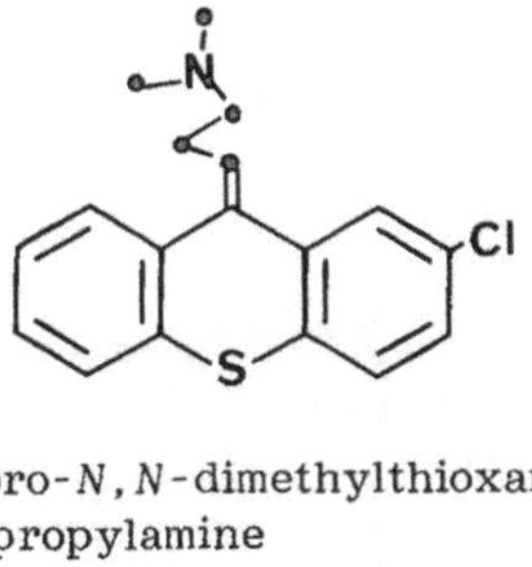

Doxepin   Chlorprothixene

Figure 27 Bioisoteres of phenothiazines possessing conformation
similar to those of phenothiazines.

Trimethobenzamine and haloperidol, two structurally quite different compounds, are also noted for their extrapyramidal syndrome. By making the molecular models of the latter compounds, one can see that these molecules can assume many different conformations, ranging from a more or less linear to a phenothiazine-like conformation. When these molecules assume the phenothiazine-like conformation, the tertiary amino group is also separated from the hydrophobic ring by a distance of about three carbon atoms (Figure 28).

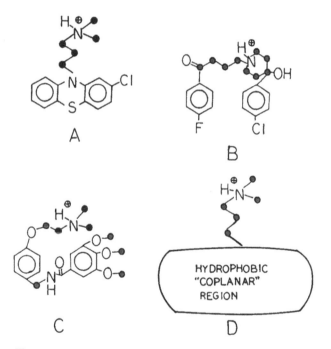

Figure 28 Common structural denominator of chlorpromazine, haloperidol and. trimethobenzamide. A, chlorpromazine; B, haloperidol; C, trimethobenzamide; D, common structural denominator of these drugs.

In clinical applications it is known that other tranquilizers, such as diazepam, do not give extrapyramidal effects. Compariosn of the molecular model of diazepam with those of phenothiazines revealed significant differences. X-ray crystallographic study showed that the phenyl group in diazepam made an angle of 124° with the chlorophenyl ring (17). In the same study it was found that diphenylhydantoin, a potent anticonvulsant, exhibited very similar structural featues with diazepam, which is a tranquilizer, a muscle relaxant, and is used in treating emergency cases of status epilepticus. The two phenyl groups and the hydantoin ring are not coplanar. These are, of course, quite different from the phenothiazines, wherein the two benzene rings are essentailly coplanar. Furthermore, neither of these two anticonvulsant drugs possesses a tertiary amino group (Figure 29).

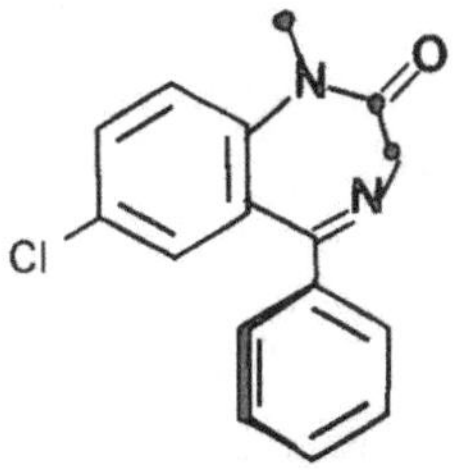

7-Chloro-1,3-dihydro-1-methyl-5-phenyl-2*H*-1,4-benzodiazepin-2-one

Diazepam

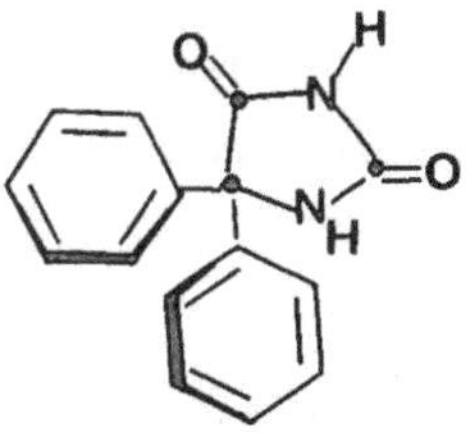

5,5-Diphenyl-2,4-imidazolidinedione

Diphenylhydantoin

Figure 29 Structural formulas of diazepam and diphenylhydantoin. Note the noncoplanarity of the ring system.

II. BONE MARROW DEPRESSION AND BLOOD DYSCRASIAS

Direct or indirect bone marrow depression may result in a spectrum of observed reduction of blood cells (e.g., agranulocytosis, leukopenia, thrombocytopenia, neutropenia, pancytopenia, and anemia). The toxic actions of drugs on the bone marrow has long been recognized by various investigators (18, 19) and are considered as the most dangerous of drug allergies (20). Extensive compilation is available on the drugs and chemicals that have been reported or implicated with blood dyscrasias (21-39).

Like skin sensitization, no single common structural denominator can be identified for the drugs reported to cause various types of hematopoietic depression, although many do fit into the categories under skin sensitization. Nevertheless, by grouping these drugs into different pharmacological groups (see Table 38), a number of interesting points appear to emerge (4).

1. Most antineoplastic agents are known to cause bone-marrow or hematopoietic depression due to their direct cytotoxicity toward rapidly dividing or growing cells as reflected by alopecia caused by many of these drugs. These drugs not only cover a wide spectrum of molecular structures but also act by many different biochemical mechanisms, such as inhibition of DNA synthesis (mitomycin, 5-fluorouracil, cytarabine, methotrexate, hydroxyurea, etc.), and inhibition of RNA synthesis (dactinomycin, daunomycin, mithramycin, adriamycin, etc.). Many antimicrobial antibiotics known to inhibit protein synthesis may also cause bone-marrow depression (gentamicin, streptomycin, lincomycin, tetracyclines, chloramphenicol, etc.). Antineoplastic agents such as alkylating agents and nitrosoureas probably act

via a multitude of mechanisms because of their high reactivity and lack of selectivity.

2. While a high degree of correlation exists between the allergic reaction and bone marrow depression for other drugs, only a relatively small percentage of antineoplastic agents has been reported to cause allergic or skin reactions. This is at least partially due to the immunosuppressive effect of these drugs.

3. Since most of the drugs listed under antimicrobial, CNS, analgetic, antipyretic, antiarthritic, anti-inflammatory, antiprotozoan, diuretic, and miscellaneous groups are known to cause both bone marrow depression and allergic or skin reactions, they probably exert the hematopoietic depression by indirect inhibition of the production of hemocytoblast and/or their differentiation into various types of cells. These adverse drug reactions are often called "idiosyncratic" drug reactions, since the majority of the population is spared from these reactions. The explanation of the idiosyncratic toxic reactions in chamical terms is difficult because of the multiplicity of biochemical reactions involved. Although most drugs known to cause bone marrow depression are also known to cause allergic or skin sensitization (except some anticancer drugs), fortunately the reverse is not true. There are many more drugs which have been reported to cause skin sensitizations or allergic reactions, but not bone marrow depression.

Among the various forms of blood dyscrasias, agranulocytosis and leukopenia appear to be most commonly reported, while only a few have been associated with pancytopenia, which results from the suppression of the stem cells. The frequency of

TABLE 38 Drugs and Chemicals Reported to Cause Blood Dyscrasias

a. Antineoplastic agents

Drug name	Agranulocytosis neutropenia leukopenia	Bone marrow depression pancytopenia	Throbocytopenia	Anemia	Allergic skin reaction
Adriamycin	X	X	X		X
Azaserine	X				
Azathiopyrine	X	X	X	X	X
BCNU	X	X	X		
Busulfan	X	X	X		
Camptothecin	X				
Carminomycin	X			X	
CCNU	X		X		
Chlorambucil	X	X	X		X
Cyclocytidine	X		X		
Cyclophosphamide	X	X			
Cytarabine (cytosine arabinoside)	X	X	X	X	X
Dactinomycin	X	X	X		X

Daunomycin	X	X			
α-2'-Deoxythioguanosine	X		X		
Dibromodulcitol	X		X		
Dimethyltriazenomidazole carboxamide	X	X	X		
Floxuridine	X		X		
5-Fluorouracil	X	X			
Hexamethylolmelamine	X				
Hydroxyurea	X	X		X	
ICRF-159	X				
Interferon	X			X	
Mechlorethamine	X	X	X	X	X
Melphalan	X	X			
6-Mercaptopurine	X	X	X	X	X
Methotrexate	X	X	X		X
4-Methyl-CCNU	X		X		
Mithramycin			X	X	
Mitomycin	X	X	X		X
Pipobroman		X			

TABLE 38 (Continued).

Drug name	Agranulocytosis neutropenia leukopenia	Bone marrow depression pancytopenia	Throbocytopenia	Anemia	Allergic skin reaction
Procarbazine	X	X			X
Streptonigrin		X			
Thioguanine	X	X			
Thio-tepa				X	
Triethylphosphamide	X	X			
Uracil mustard	X				
Urethane		X			
Vinblastine	X	X	X		
Vincristine	X	X	X		

b. Antimicrobial agents

Drug name	Agranulocytosis neutropenia leukopenia	Bone marrow depression pancytopenia	Throbocytopenia	Anemia	Allergic skin reaction
Aminosalicylates	X		X	X	X
Amphotericin B	X	X	X	X	X

	1	2	3	4	5
Ampicillin			X	X	X
Azlocillin	X			X	X
Cephaloridine	X			X	
Cephalosporins	X			X	X
Cephalothin	X				X
Chloramphenicol	X	X	X	X	X
Clindamycin				X	X
Cloxacillin	X				X
Colistin	X		X	X	X
Co-trimoxazole (trimethoprim + sulfamethoxazole)	X		X	X	X
Dapsone	X		X	X	X
Flucytosine	X	X			
Fumagillin	X				X
Gentamycin	X		X		
Isoniazid	X		X	X	X
Ketoconazide				X	X
Lincomycin	X				X
Merbromin				X	

TABLE 38 (Continued).

Drug name	Agranulocytosis neutropenia leukopenia	Bone marrow depression pancytopenia	Throbocytopenia	Anemia	Allergic skin reaction
Methicillin	X	X	X	X	X
Miconazole	X			X	X
Nafcillin	X				
Nalidixic acid				X	X
Nitrofurantoin	X	X	X	X	X
Novobiocine	X				X
Penicillin G				X	X
Penicillin V				X	X
Povidone-iodine	X				
Rifampicin			X	X	
Ristocetin	X		X		X
Salicylazosulfapyridine	X			X	X
Streptomycin sulfate	X		X	X	
Sulfachlorpyridine	X				
Sulfadiazine silver	X				X

	Agranulocytosis neutropenia leukopenia	Bone marrow depression pancytopenia	Throbocytopenia	Anemia	Allergic skin reaction
Sulfaguanidine	X				
Sulfameter	X				X
Sulfamethoxypyridazine	X				
Sulfapyridine	X				
Sulfasalazine	X				
Sulfathiazole	X				
Sulfonamides	X	X	X	X	X
Sulfones	X				
Tetracyclines	X				

c. CNS drugs, sedative-hypnotics, anticonvulsants, neuroleptics, and antidepressants

Drug name	Agranulocytosis neutropenia leukopenia	Bone marrow depression pancytopenia	Throbocytopenia	Anemia	Allergic skin reaction
Carbamazepine	X		X	X	X
Clozapine	X			X	
Ethosuximide	X		X	X	X
Glutethimide	X			X	X
Heptabarbital	X		X	X	X
Mephenytoin	X		X	X	X

TABLE 38 (Continued).

Drug name	Agranulocytosis neutropenia leukopenia	Bone marrow depression pancytopenia	Throbocytopenia	Anemia	Allergic skin reaction
Methsuximide	X				X
Nitrous oxide		X			
Phenacemide	X				X
Primaclone (primidone)	X		X		
Sulthiame	X				
Trimethadione	X	X			X
Vaproate sodium		X	X		
Acetophenazine	X			X	X
Butaperazine	X				X
Chordiazepoxide	X				X
Chlorpromazine	X		X	X	X
Chlorprothixene	X		X	X	X
Fluphenazine	X				X
Haloperidol	X			X	
Lithum carbonate	X			X	

Mephenoxalone	X				X
Meprobamate	X	X	X		X
Mesoridazine	X				X
Methotrimeprazine					X
Oxazepam	X				X
Perazine					X
Perphenazine					X
Phenothiazines	X	X	X		X
Piperacetazine	X		X		X
Prochlorperazine	X				X
Promazine	X				X
Thioridazine	X		X		X
Thiothixene	X		X		X
Amitriptyline	X				X
Desipramine	X				X
Imipramine	X				X
Isocarboxazide	X	X			
Mianserine		X			X
Nomifensine				X	

TABLE 38 (Continued).

Drug name	Agranulocytosis neutropenia leukopenia	Bone marrow depression pancytopenia	Throbocytopenia	Anemia	Allergic skin reaction
Nortriptyline	X				X
Phenelzine	X				X

d. Anti-inflammatory, analgetic, antiarthritic, and uricosuric agents

Drug name	Agranulocytosis neutropenia leukopenia	Bone marrow depression pancytopenia	Throbocytopenia	Anemia	Allergic skin reaction
Acetaminophen	X				X
Acetylsalicylic acid				X	X
Allopurinol	X	X	X		X
Aminopyrine	X	X	X	X	X
Antrafenine			X		
Auranofin			X	X	X
Aurothiopolypeptide	X				
Aurothioglucose	X		X	X	X
Benoxaprofen			X		

Clometacin	X		X		
Colchicine	X	X	X	X	X
Dextropropoxyphen	X				
Diclofenac	X		X	X	
Dipyrone	X	X	X		X
Fenoprofen	X		X	X	
Gold sodium thiomalate			X		X
Ibuprofen	X	X	X	X	
Indomethacin	X	X	X	X	X
Mefenamic acid	X	X	X	X	
Methapyrone	X				X
Naproxen	X	X	X		
Niflumic acid	X				
Oxyphenbutazone	X	X	X		
Pentazocin	X	X	X	X	X
Phenacetin	X	X	X	X	X
Phenylbutazone	X	X	X	X	X
Probenecid	X	X	X	X	
Pyritinol	X		X		X

TABLE 38 (Continued).

Drug name	Agranulocytosis neutropenia leukopenia	Bone marrow depression pancytopenia	Throbocytopenia	Anemia	Allergic skin reaction
Salicylates			X	X	X
Sulfinpyrazone	X		X	X	
Sulindac	X	X	X	X	X
Tolmetin	X		X	X	

e. Antiprotozoan, diuretic, and miscellaneous drugs

Drug name	Agranulocytosis neutropenia leukopenia	Bone marrow depression pancytopenia	Throbocytopenia	Anemia	Allergic skin reaction
Acetyldigoxin			X		
Alfafa seeds		X		X	
Amodiaquin	X	X		X	
Amrinone			X		
Antazoline			X		
Antimony potassium tartrate			X		

Antithymocyte globulin				X
Chlorguanide	X			
Chloroquin	X	X		X
Cimetidine	X	X	X	X
Hydroxychloroquin	X	X	X	X
Primaquin	X		X	
Quinacrine hydrochloride	X		X	X
Quinine	X	X		X
Stibophen		X	X	X
Suramin sodium	X			X
Acetazolamide	X	X		
Benzoflumethiazide	X	X	X	X
Benzothiazides (hydrochlorothiazides)	X	X		X
Captopril	X			X
Chlorthalidone	X	X		
Dichlorphenamide	X	X		
Enalapril	X(?)		X(?)	
Ethacrynic acid	X	X	X	X

TABLE 38 (Continued).

Drug name	Agranulocytosis neutropenia leukopenia	Bone marrow depression pancytopenia	Throbocytopenia	Anemia	Allergic skin reaction
Ethoxyzolamide	X	X	X		X
Furosemide	X	X	X	X	X
Mercurial diuretics	X	X			X
Methazolamide	X	X	X		X
Metolazone		X		X	
Thiazides	X			X	
Trichlormethazide	X				X
Acetohexamide	X				X
Chlorpropamide	X		X		X
Clofibrate	X				X
Coumarin compounds	X				X
Diatrizoate	X				X
Digoxin	X				X
Doxepin	X				
Edetates		X		X	X

Ethopropazine	X			X	X
Heparin			X		
Hydralazine	X	X	X	X	X
Hydroxyamphetamine	X		X	X	X
Idoxuridine	X		X		
Levamisole	X				X
Levodopa	X			X	X
Mebendazole	X				
Mebhydrolin	X				
Metiamide	X				
Methimazole	X				X
Methyldopa		X	X	X	X
Methylsergide maleate	X		X	X	X
Metronidazole	X	X		X	
Noramidopyrine	X	X			
Orphenazine				X	
Oxyprenolol		X			
Penicillamine	X	X	X	X	X
Phenindione	X				X

TABLE 38 (Continued).

Drug name	Agranulocytosis neutropenia leukopenia	Bone marrow depression pancytopenia	Throbocytopenia	Anemia	Allergic skin reaction
Potassium perchlorate	X				
Procainamide HCl	X	X	X	X	X
Promethazine	X	X	X	X	X
Propylthiouracil	X	X	X	X	X
Quinidine	X	X	X	X	X
Ritodine		X		X	X
Sulfonylureas	X		X		X
Thenoldine	X				
Thorium dioxide	X	X			
Ticlopidine	X	X	X		X
Tocainide				X	X
Tolazamide	X	X	X		
Tolbutamide	X	X	X	X	X
Trimeprazine	X	X	X	X	
Tripelennamine	X	X		X	
Vitamin K		X		X	

Source: From Refs. 4, 18—39, and 41, and M. N. G. Dukes and J. Elias, Eds., 1980—1984, Side Effects of Drugs Annual 4—8, Excerpta Medica, Amsterdam.

thrombocytopenia and anemia is somewhere in between. There appears to be a relationship between the half-lives of the cell lines and the tendency of the specific cell line to be suppressed. For example, the half-life of granulocytes is about 6 hr, while the half-life of platelets is about 5 to 7 days; these two are more frequently involved in earlier suppression by various drugs. On the other hand, red cells have half-lives of about 120 days, and anemia occurs much slower than neutropenia or agranulocytosis. While thrombocytopenia is rarely the dominant limiting side effect, platelets are the slowest to recover after bone marrow suppression (40).

At present, no apparent structural features or physico-chemical properties can be associated with the different patterns of hematopoietic depression. Whether one can unravel this complicated matrix of information by using other techniques such as pattern recognition or cluster analysis remains to be explored.

III. HEPATOTOXICITY

Many lipophilic drugs, poisons, food additives, and environmental pollutants, after entry into the body through various routes, have a tendency to accumulate in the lipoid tissues and/or to remain bound to plasma and tissue proteins. Most of them can only be cleared from the body after biotranformations into more hydrophilic metabolites. These biotransformations are catalyzed by various enzyme systems located predomininantly in the liver (5):

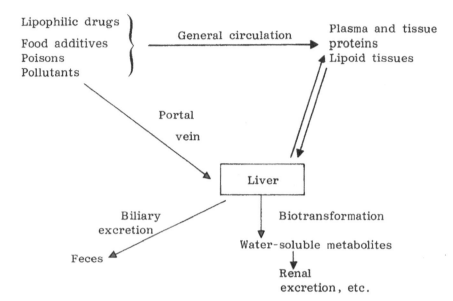

Since the liver is one of the most important organs involved in many important biotransformations of drugs and endogenous substances, it is exposed to many active metabolites as well as to the parent molecules. This is due to a profuse supply of blood and the presence of many redox systems, such as cytochromes and various enzymes involved in the formation of bile pigments. In a five-year study by the Swedish Adverse Drug Reaction Committee, 578 (21%) of 2750 reported cases were classified as hepatocellular damage (42).

One can classify hepatotoxins according to symptoms, mechanisms of action, and experimental criteria. The following classification is mainly for the sake of convenience and by no means absolute, since overlaps do occur (5, 20).

A. Direct Hepatotoxins

These chemicals have a short latent period (hours to days). The degree of liver damage is dose dependent and the results are experimentally reproducible in animals. All heavy metals and many cytotoxic antineoplastic agents belong to this group. Some anticancer drugs are not listed as hepatotoxic. This is probably due to greater sensitivies of other organs and tissues. For example, the limiting toxicity of adriamycin is cardiomyopathy, and hydroxyurea and 5-fluorouracil can cause leukopenia and bone marrow depression before signs of hepatotoxicity are observed. Alcohol and halogenated hydrocarbons such as chloroform and carbon tetrachloride are considered as direct hepatotoxins, although bioactivation is probably involved in the process.

B. Drugs Causing Hepatotoxicity via Bioactivation and Covalent Binding (Acute Necrotic or Hypersensitivity Type)

Many highly reactive intermediates, such as arene oxide, epoxide, and iminoquinone and free radicals, have been reported to be responsible for the cell damage or death following activation by cytochrome P-450 or other redox enzymes. For example, guanoxan, an antihypertensive drug, has been shown to be hepatotoxic. Since 9.2 to 53.6% of the phenolic metabolites can be found in a 24-hr urine sample (43), it is very likely that the arene oxide is the reactive metabolite responsible for the hepatotoxicity (Scheme 24).

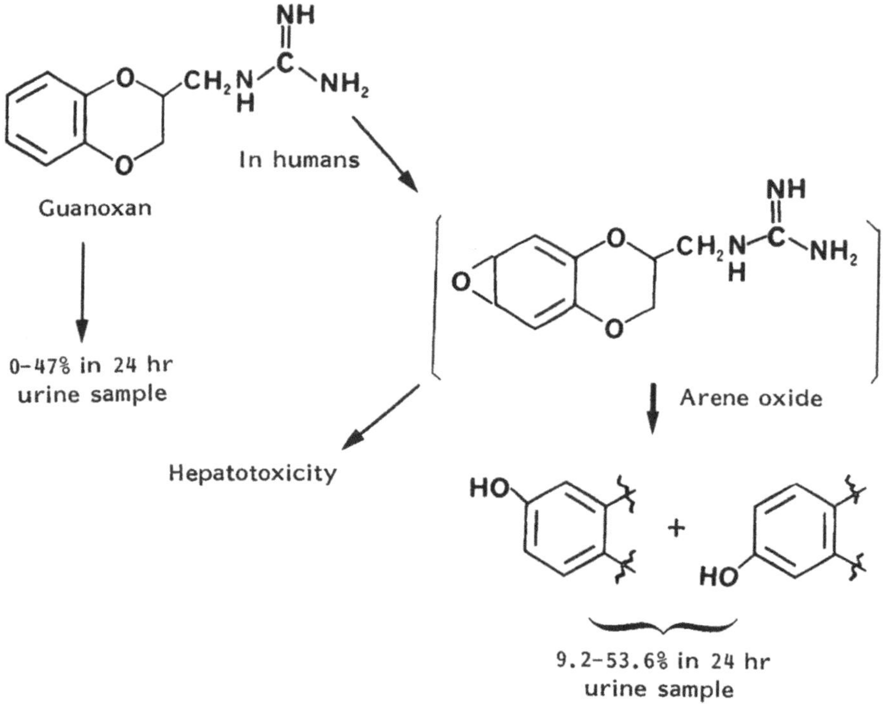

Scheme 24

Ticrynafen (Selacryn) marketed by Smith, Kline & French in 1979 as an antihypertensive and diuretic was removed from the market due to its hepatotoxicity observed in patients. From the structural point of view, this may be due to epoxide formation of the thiophene ring carrying a C=O group (36):

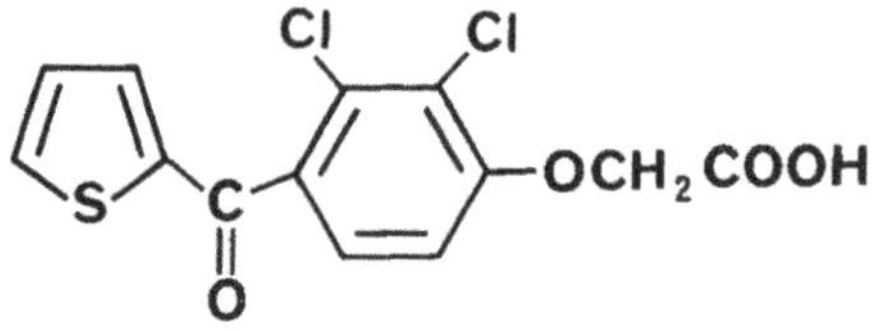

(36)

Ticrynafen (Selacryn)

Other drugs capable of covalent bond formation through reactive intermediates include isoniazid, iproniazid, acetaminophen phenothiazines, carbamazepine, sulfonamides, sulfonylureas, and α-methyldopa. Some drugs are capable of direct covalent bond formation with nucleophiles (e.g., penicillins, cephalosporins, griseofulvin, ethacrynic acid). Many of these drugs are also known to cause skin sensitization, hypersensitivity and eosinophilia (3—5).

C. Drugs Interfering with the Binary Uptake, Secretory, or Excretory Pathways of Bile

Acidic compounds such as flavaspidic acid (male fern oil), buniodyl, and iodipamide (contrast media) can compete with bilirubin for the binding of Z-transport protein and result in jaundice (5, 20). Novobiocin has been reported to inhibit uridine diphosphate glucuronyl transferase—thus decreases bilirubin conjugation and results in jaundice. Low incidence of simple cholestasis has been associated with anabolic hormones (1 to 2%) and oral contraceptives. Oral contraceptive-induced cholestasis and jaundice appear to be most frequent in the Scandinavian countries and Chile, suggesting that a certain genetic factor is involved (44). Benoxaprofen (Oraflex) was removed from the market shortly after it was introduced, due to death following onset of cholestatic jaundice in elderly patients (45).

D. Drugs Capable of Inducing Hepatitis

Hepatitis induced by chemically fairly inert general anesthetics such as halothane and methoxyflurane has been well documented

TABLE 39 Drugs Reported to Cause Various Types of Hepatotoxicities

Drug	Hepatitis	Hepato-toxicity (liver damage, function abnormal)	Cholestatic jaundice (obstruc-tive, cholestasis)	Hepato-cellular jaundice	Unspecified jaundice	Liver necrosis	Liver cirrhosis	Hepatic coma	Hepato-megaly (liver enlarge-ment)	Hepatic fibrosis
Acetaminophen (paracetamol)	X	X			X	X				
Acetarsol					X					
Acetazolamide								X		
Acetohexamide			X	X	X					
Acetophenazine		X			X					
Acetylsalicylate acid	X	X		X						
Acetylureas	X									
Ajmaline			X							
Aldrin	X									
Alkylating agents		X								
Allopurinol	X	X	X	X						
Aloudine	X									
Alvestatin		X								
Althesin		X								
Aluminum		X								

214

Amfenac		X						
Amidon	X							
Amidopyrine		X						
Amineptine		X						
Aminocaproil acid				X		X		
Aminophenazone		X	X					
Aminosalicylates		X						
Amiodarone		X						X
Amithiazone	X							
Amitriptyline	X		X		X			
Ammonium chloride							X	
Amodiaquin	X	X						X
Amphetamine	X	X						
Amphotericin B		X			X		X	
Ampicillin		X						
Amsacrine		X					X	
Anabolic agents		X	X					
Androgens			X					
Anileridine	X							
Anticonvulsants	X	X		X	X			
Antileprosy drugs	X							

TABLE 39 (Continued).

Drug	Hepatitis	Hepato-toxicity (liver damage, function abnormal)	Cholestatic jaundice (obstructive, cholestasis)	Hepato-cellular jaundice	Unspecified jaundice	Liver necrosis	Liver cirrhosis	Hepatic coma	Hepato-megaly (liver enlarge-ment)	Hepatic fibrosis
Antimetabolites	X	X								X
Antimony potassium tartrate	X	X			X					
Antimony preparations					X	X				
Antirheumatic drugs	X									
Arsenical drugs	X	X								
Arsphenamine (salvarsan)			X							
L-Asparaginase		X								
Azacytidine		X								
Azathioprime	X	X	X	X						
Azetepa		X								
Baclofen		X								
Barbiturates	X	X								
Barium sulfate		X								
BCG	X									

The following is a portion of an index table (adverse-effect cross-reference). The column headings are not present on this page; the marks (X) are reproduced in their apparent columns.

Drug	1	2	3	4	5	6	7
Benemid	X						
Benoxaprofen			X				
Benorylate	X						
Benzedrine	X						
Benziodarone					X		
Benzomorphans	X						X
Benzothiadiazines		X					
Benzothiazides			X				
β-Phenylisopropyl-hydrazine						X	
Bismuth compounds		X			X		
1,3-Bis(2-chloroethyl)-1-nitrosoureas		X					
Bromocriptine	X(?)	X(?)			X(?)		
Bunamiodyl	X			X			
Capreomycin	X						
Captopril	X						
Carbamazepine	X	X	X	X		X	
Carbasone	X		X	X			
Carbenoxolone	X	X	X				
Carbimazole	X			X		X	
Carbon tetra-chloride	X			X		X	

217

TABLE 39 (Continued).

Drug	Hepatitis	Hepatotoxicity (liver damage, function abnormal)	Cholestatic jaundice (obstructive, cholestasis)	Hepatocellular jaundice	Unspecified jaundice	Liver necrosis	Liver cirrhosis	Hepatic coma	Hepatomegaly (liver enlargement)	Hepatic fibrosis
Carbutamide				X						
Carprofen		X								
Cefaperazone		X								
Cephalexin		X								
Cephaloridine		X								
Cephalosporins		X								
Cephalothin sodium		X								
Chenodeoxycholic acid	X	X								
Chloroform	X	X		X	X	X				
Chloral hydrate		X								
Chlorambucil		X	X	X						
Chloramphenicol		X	X		X					
Chlordiazepoxide	X		X	X	X					
1-(2-Chloroethyl)-3-cyclohexyl-1-nitrosourea		X								

	1	2	3	4	5	6
Chlorothiazide		X	X		X	
Chlorozotocin						X
Chlormezanone					X	
Chlorpromazine			X	X	X	X
Chlorpropamide		X	X	X	X	X
Chlorprothixene					X	
Chlortetracycline					X	
Chlorzoxazone				X		
Cholera vaccine			X			
Cholestyramine				X		
Cimetidine					X	X
Cincophen				X		X
Clindamycin						X
Clofibrate						X
Clometacin			X			
Clorazepate dipotassium				X		
Clotrimazole				X		X
Clozapine						X(?)
Colchicine						X
Comfrey (Symphytum officinale)				X		X
Copper sulfate						

TABLE 39 (Continued).

Drug	Hepatitis	Hepato-toxicity (liver damage, function abnormal)	Cholestatic jaundice (obstructive, cholestasis)	Hepato-cellular jaundice	Unspecified jaundice	Liver necrosis	Liver cirrhosis	Hepatic coma	Hepato-megaly (liver enlarge-ment)	Hepatic fibrosis
Coumarin derivatives		X								
Crotalaria		X		X						
Cyclophosphamide		X		X	X					
Cytarabine		X								
Dacarbazine (DTIC; DIC)		X								
Danazol		X			X					
Dantrolene	X	X								
Dapsone	X	X		X						
Deferoxamine		X								
Desacetyl-methylcolchicine	X		X							
Desipramine			X		X					
Dextropropoxy-phene		X								
Dextrothyroxine		X								
Diacetylmorphine	X	X								

Compound	1	2	3	4	5	6
Diazepam	X	X				X
Diclofenac	X	X				
Diethylamino-ethoxyhexoestrol	X	X				
Di-2-ethyl-phthalate (DEHP)	X	X				
Diethylstilbestrol	X	X	X			
Difetarsone		X				
Diflurisal		X				
Dinitrophenol		X				
Diphenylhydantoin	X	X		X		
Disopyramide		X		X		
Disulfiram	X	X				
Doxycycline		X				
Econazole	X					
Ectylurea			X		X	
Enflurane	X	X				
Erythromycin		X	X		X	
Erythromycin estolate	X	X	X		X	
Estradiol					X	
Estriol					X	
Estrogens		X				

TABLE 39 (Continued).

Drug	Hepatitis	Hepato-toxicity (liver damage, function abnormal)	Cholestatic jaundice (obstruc-tive, cholestasis)	Hepato-cellular jaundice	Unspecified jaundice	Liver necrosis	Liver cirrhosis	Hepatic coma	Hepato-megaly (liver enlarge-ment)	Hepatic fibrosis
Ethacrynic acid		X								
Ethambutol		X								
Ethanol	X	X				X	X			
Ethchlorvynol	X	X								
Ethenyloestranol			X							
Ether (ethyl)	X			X						
Ethinamate	X									
Ethionamide	X	X		X						
Ethotoin				X						
Ethylchloride		X								
Ethylestrenol		X	X		X					
Etretinate	X	X								
Fat emulsions		X				X				
Fenbufen		X								
Fenofibrate	X	X								
Fenoxypropazine		X				X				
Fentiazac		X								

	1	2	3	4	5	6	7
Feprazone		X		X			
Ferrous sulfate		X					
Flucloxacillin		X					
Flucytosine		X					
Fluorescin		X					
Fluphenazine		X	X		X		
Flurothyl			X				
Fluroxene		X					
Fluoxymesterone		X	X		X		
Furosemide					X	X	X
Fusidic acid sodium		X			X		
General anesthetics		X					
Glafenine	X	X					
Glibenclamide	X					X	
Glycerol		X					
Gold compounds	X	X		X	X		
Gold thiomalate sodium		X			X		
Griseofulvin	X	X	X	X			
Guanoxan		X			X		
Halothane	X	X		X	X		
Haloperidol		X					

TABLE 39 (Continued).

Drug	Hepatitis	Hepato-toxicity (liver damage, function abnormal)	Cholestatic jaundice (obstructive, cholestasis)	Hepato-cellular jaundice	Unspecified jaundice	Liver necrosis	Liver cirrhosis	Hepatic coma	Hepato-megaly (liver enlarge-ment)	Hepatic fibrosis
Heparinoids		X								
Heptachlor		X								
Hexachloro-paraxylol					X					
Hexachlorothane	X	X								
Human leukocyte interferon		X								
Hycanthone mesylate		X				X				
Hydantoins	X	X							X	
Hydralazine	X	X	X	X						
Hydrochloro-thiazide					X					
Hydroxystil-bamide		X								
Ibufenac		X			X					
Ibuprofen		X								
Idoxuridine		X								

Drug							
Imipramine	X	X		X			
Immunosuppressants				X			
Indandione derivatives	X			X			
Indomethacin	X	X	X	X			
Iodine compounds				X			
Iodipamide	X		X				
Iopanoic acid	X		X				
Iprindole		X					
Iproniazid	X		X	X	X		X
Iron (overload)	X					X	
Isaxozine	X						
Isocarboxazid			X	X	X		
Isofluorphate (Dyflos)				X			
Isoflurane	X						
Isoniazid	X			X			
Isophosphamide	X						
Isoxepac	X						
Josamycin	X						
Ketokonazole	X			X	X		
Ketophenyl-butazone	X			X			

TABLE 39 (Continued).

Drug	Hepatitis	Hepato-toxicity (liver damage, function abnormal)	Cholestatic jaundice (obstruc-tive, cholestasis)	Hepato-cellular jaundice	Unspecified jaundice	Liver necrosis	Liver cirrhosis	Hepatic coma	Hepato-megaly (liver enlarge-ment)	Hepatic fibrosis
Lentinam		X								
Lergotrile	X									
Levodopa		X								
Lincomycin					X					
Lorazepam		X								
Lonazolac	X									
Macrolide anti-biotics	X	X			X					
MAO inhibitors	X	X		X						
Meglumine iodip-amide		X								
Melarsoprol		X								
Mepazine			X							
Mephenytoin				X						
Meprobamate			X							
Mercaptopurine	X	X	X	X	X					
Mercury		X								

	1	2	3	4	5	6	7	8	9
Metahexamide			X	X					
Methamphetamine	X								
Methandrostenolone		X	X						
Methimazole		X	X	X					
Methisazone		X							
Methotrexate		X	X	X					
Methoxsalen		X	X	X					
Methoxyflurane	X	X	X	X			X		X
Methyldopa	X	X	X		X				
Methylestrenolone			X			X		X	
Methylpheniate		X							
Methyltestosterone		X	X		X				
Mithramycin		X		X					
Mono-octanoin		X							
Monophenylbutazone		X							
Morphazinamide					X				
N-Methylformamide		X							
Naldixic acid	X		X						
Naproxen	X(?)	X	X	X					

TABLE 39 (Continued).

Drug	Hepatitis	Hepato-toxicity (liver damage, function abnormal)	Cholestatic jaundice (obstructive, cholestasis)	Hepato-cellular jaundice	Unspecified jaundice	Liver necrosis	Liver cirrhosis	Hepatic coma	Hepato-megaly (liver enlargement)	Hepatic fibrosis
Neocincophen	X									
Niacin (nicotinic acid)		X	X		X					
Nialamide	X	X			X					
Nicotinamide				X		X				
Nicotinyl alcohol		X								
Nifedipine	X	X			X					
Niridazole		X								
Nitrofurans		X								
Nitrofurantoin	X	X	X							
Nitrosoureas		X								
Nominfensine	X									
Norethandrolone		X	X							
Norethindrone			X							
Nortriptyline			X							
Novobiocin		X		X						
Oleandomycin		X			X					

	1	2	3	4	5	6	7	8
Opioid analgesics	X							
Oral contraceptives	X	X	X			X		
Oxacillin	X	X	X					
Oxamniquine		X						
Oxandrolone		X	X					
Oxazepam		X						
Oxophenarsine		X						
Oxymetholone		X	X					
Oxyphenbutazone	X			X				
Oxyphenisatin acetate	X	X	X	X	X			
p-Aminosalicylic acid (PAS)	X	X		X				
Papaverine		X						
Pemoline		X						
Penicillamine	X	X	X					
Penicillins		X		X				
Pentazocine								X
Perhexiline	X	X					X	
Perphenazine			X					
Pertussive vaccine		X						
Phenacemide	X	X		X				

TABLE 39 (Continued).

Drug	Hepatitis	Hepato-toxicity (liver damage, function abnormal)	Cholestatic jaundice (obstructive, cholestasis)	Hepato-cellular jaundice	Unspecified jaundice	Liver necrosis	Liver cirrhosis	Hepatic coma	Hepato-megaly (liver enlargement)	Hepatic fibrosis
Phenacetin		X	X							
Phenelzine	X	X			X	X				
Phenidione	X	X	X		X	X				
Pheniprazine	X	X				X		X		
Phenobarbital		X		X						
Phenoperidine		X								
Phenophthalein				X						
Phenothiazines	X	X	X		X					
Phenylbutazone	X	X	X	X	X	X				
Phenylcinchonic acid						X				
Phenytoin	X	X								
Pirprofen	X	X								
Piroxicam	X(?)									
Polyvinyl chloride (plasticizer)	X	X								

230

Drug	1	2	3	4	5	6
Providone						X
Praziquantel	X					
Prednisolone	X (in hepatitis B patients)					
Probenicid	X		X		X	
Procarbazine	X			X		
Prochlorperazine		X				
Progestans	X					
Promazine		X				
Promethazine		X				
Propylthiouracil	X	X	X			
Prothionamide	X					
Psoralens	X					
Pyrazinamide	X		X	X	X	
Pyrazinamide-isoniazid	X		X	X	X	
Pyrazolone derivatives	X			X		
Pyrrolizidine alkaloids	X					
Quinacrine hydro-chloride	X		X			
Quinethazone		X				
Quinidine	X					
Ranitidine	X(?)					

TABLE 39 (Continued).

Drug	Hepatitis	Hepato-toxicity (liver damage, function abnormal)	Cholestatic jaundice (obstructive, cholestasis)	Hepato-cellular jaundice	Unspecified jaundice	Liver necrosis	Liver cirrhosis	Hepatic coma	Hepato-megaly (liver enlargement)	Hepatic fibrosis
Rifampicin		X								
Rifamycin		X								
Rufocromomycin		X								
Salicylates		X								
Saramycetin		X								
Seatone	X									
Selenum		X								
Stanozolol		X	X							
Stibocaptate		X								
Stibophen	X									
Sudoxicam		X								
Sulfadiazine			X							
Sulfamethoxazole		X								
Sulfamethoxy-pyridazine						X				
Sulfanilamide			X							
Sulfonamides	X	X	X	X	X					

	1	2	3	4	5	6	7	8
Sulfones	X	X						
Sulfonylureas	X	X	X					
Sulindac		X			X		X	
Tannic acid		X						
Tetracyclines		X	X					
Thiabendazole		X						
Thiacetazone		X			X			
Thiamazole			X					
Thiazide drugs		X	X			X		
Thiazine diuretics						X		
Thiocarlide					X			
Thioguanine	X		X		X			
Thiopental					X			
Thioridazine	X	X	X					
Thiothixene		X						
Thiouracil			X	X				
Thorium dioxide								X
Ticlopidine			X					
Ticrynafen	X	X						
Tiopronin		X						
Tolazamide					X			
Tolbutamide		X	X					

TABLE 39 (Continued).

Drug	Hepatitis	Hepato-toxicity (liver damage, function abnormal)	Cholestatic jaundice (obstructive, cholestasis)	Hepato-cellular jaundice	Unspecified jaundice	Liver necrosis	Liver cirrhosis	Hepatic coma	Hepato-megaly (liver enlarge-ment)	Hepatic fibrosis
Tolfenamic acid	X									
Total intravenous nutrition			X							
Tranylcypromine		X								
Triacetyloleando-mycin	X	X		X						
Triacetylphen-isatin	X									
Triaziquone			X							
Tribromethanol		X								
Trifluperazine			X							
Timeprazine			X							
Trimethadione	X	X		X						
Tromethamine (TRIS, THAM)		X								
Trimethobenz-amide	X									
Trimethoprim-sulfamethoxazole		X								

Triparanol		X								
Tripelenamine			X							
Trithiozine	X									
Troleandromycin		X			X					
Tryparsamide		X								
Uracil mustard		X			X					
Urethane		X		X		X		X		X
Ursodeoxycholic acid		X								
Vaproute sodium	X	X					X			
Vasopressors						X				
Veramon			X							
Verapamil	X	X							X	
Vinblastine										X
Vincristine		X								
Vinyl ether		X		X						
Viscum album (mistletoe)	X				X					
Vitamin A (overdose)				X			X		X	
Vitamin K (in infants)					X					

TABLE 39 (Continued).

Drug	Hepatitis	Hepato-toxicity (liver damage, function abnormal)	Cholestatic jaundice (obstructive, cholestasis)	Hepato-cellular jaundice	Unspecified jaundice	Liver necrosis	Liver cirrhosis	Hepatic coma	Hepato-megaly (liver enlarge-ment)	Hepatic fibrosis
Yellow fever vaccine	X	X								
Zoxazolamine	X			X		X				

Source: Refs. 5, 20, and 44–52 and M. N. G. Dukes and J. Elis, Eds., 1980–1984, *Side Effects of Drugs Annual 4–8*, Excerpta Medica, Amsterdam.

236

and reviewed (46). It is now generally believed that these
fairly stable drugs cause hepatotoxicity by a free radical mechanism (5). Some drugs have been reported to cause histological,
clinical, and laboratory manifestations closely resembling those
of icteric viral hepatitis and may even cause death. These include cincophen, iproniazid, and other monoamine oxidase inhibitors, such as pheniprazine, nialamid, and phenelzine; (47)
muscle relaxants such as zoxazolamine; antituberculosis drugs
such as the pyrazinamide-isoniazid combination; (48) and indo-
methain (49). Some of these have already been withdrawn from
the market. Other drugs may cause similar reactions but
without any reported fatalities. These include isoniazid, p-
aminosalicylate (50, 51), α-methyldopa (51), chlordiazepoxide,
diazepam, gold salts, benemid (5), and thionamide (52). Table
39 summarized various drugs reported to cause different types
of hepatotoxicities.

IV. NEPHROTOXICITY

Being the major organs for disposing of wastes, drugs, and
poisons, the kidneys receive about 25% of the cardiac output
every minute, even though they make up only 0.4% of body
weight (53). The high rate of blood circulation plus the counter-
current effect in the hairpin of the loop of Henle (54) during
the process of urine formation make the kidneys particularly
vulnerable to many drugs and poisons by various biochemical
or even physical mechanisms. To complicate the situation further,
many patients receiving drugs such as antibiotics or anticancer
agents may already have impaired renal functions, which make

them even more susceptible to the toxic effects of these drugs (55–57). It is not surprising that there are many reports in the literature describing the drug-induced or drug-related nephrotoxicities in humans as well as in experimental animals (6).

A literature survey has been done on drug- and chemical-associated nephrotoxicities. Among the 1450 entries examined, about 225 (15.5%) have been associated with various types of renal toxicities. These include heavy metals, halogenated organic solvents and general anaesthetics, glycols, analgesics, and nonsteroidal anti-inflammatory agents (prostaglandin synthetase inhibitors), sulfonamides, various antibiotics (aminoglycosides, penicillins and cephalosporins, tetracyclines, amphotericin B, polymyxins, and vancomycin), radiographic contrast media, and others (6). Since then over 50 drugs have been added to the list (see Table 40).

A. Analgesic and Nonsteroidal Anti-inflammatory Drugs

Analgesic Associate Nephropathy (AAN) was first reported by Zollinger and Spühler (58) in Switzerland. The condition was later found to be worldwide, especially in Australia (59–62). Various types of renal toxicities from nephritis to papillary necrosis have been attributed to aspirin (used along and in combination), acetophenetidine (in combination), acetaminophen (alone and in combination), phenylbutazone, oxyphenbutazone, ketophenbutazone, sulfinpyrazone, indomethacin, ibuprofen, ketoprofen, fenoprofen, tolmetin, glaphenine, mefenamic acid, bucloxic acid, niflumic acid, aclofenac, naproxen, and others (61).

Several factors and mechanisms have been proposed for AAN (61): (a) Direct toxic effect by the high concentrations in the medullary tissue (loop of Henle). (b) Anoxia caused by vasoconstriction or mesangial thickening, platelet aggregates due to inhibition of prostaglandin synthesis, occlusion of blood vessels by interstitial hyperplasia, changes in the oxygen binding of hemoglobin, and changes in viscosity. (c) Metabolic effects such as reduced intracellular ATP and glucogenesis rate (63) and uncoupling of oxidative phosphorylation by salicylates. (d) Inhibtion of prostaglandin (PG) biosynthesis. Since the renal papilla is the main source of prostaglandins in the kidney, nonsteroidal anti-inframmatory agents such as aspirin and indomethacin may have deleterious effect in patients with underlying renal disease (64). It is known that all the nonsteroid anti-inflammatory agents can produce renal papillary necrosis in rats under experimental conditions (59).

When the synthesis of PG is inhibited, the effects of autonomic neurotransmitters, such as renin-angiotensin and noradrenalin, will be affected. PGE_2 is also known to produce vasoconstriction in the isolated perfused kidney (61). Both aspirin and indomethacin can produce sodium and water retention, presumably through activation of the renin-angiotension system as a consequence of the inhibition of PG synthesis (53, 65). An immunological basis has been suggested, although conflicting data appear to exist (66, 67).

Caffeine is present in many analgesic mixtures and can undoubtedly contribute to abuse tendency for its CNS stimulation effect. It has been suggested on epidemiological ground that caffeine should be as suspect as phenacetin (68). There has also been direct evidence of renal toxicity caused by caffeine

TABLE 40 Drugs and Chemicals Known to Cause Various Types of Nephrotoxicities

Drug or chemical	Nephritis Glomerulo Interstitial Pyelo-	Kidney function abnormal Kidney damage Nephropathy Nephrosclerosis Nephrosis Nephrotoxicity Renal failure Renal insufficiency	Kidney necrosis Renal papillary necrosis tubula necrosis	Nephrocalcinosis Renal calculi Renal stones	Anuria dysuria
Acetaminophen (paracetamol)	X	X	X		X
Acetazolamide		X	X		
Acetaphenazine		X			
Acetophenetidin (phenacetin)	X	X	X	X	X
Acetylpenicillamine		X			
Acetylsalicylic acid (aspirin)		X	X	X	
Acyclovia		X			
Alclofenac	X	X			
Allopurinol	X	X	X		
Amantadine			X	X	

240

Amikacin		X			
Aminocaproic acid		X			
Aminopyrine (amido-pyrine)		X			X
p-Aminosalicylic acid	X	X	X		
Amoxapine (over-dose)		X			
Amphotericin B	X	X		X	X
Antimony compounds	X	X			X
Antipyrine	X	X	X		X
Antrafenine		X			
Arsenical compounds	X	X	X		
Ascorbic acid	X	X			
Asparaginase		X			
Auranofin		X			
Auriothioglucose (also see gold compounds)	X	X			
Azacitidine		X			
Azapropazone	X				
Azathioprine	X	X			
Bacitracin	X	X	X		
BCNU		X			

TABLE 40 (Continued).

Drug or chemical	Nephritis Glomerulo Interstitial Pyelo-	Kidney function abnormal Kidney damage Nephropathy Nephrosclerosis Nephrosis Nephrotoxicity Renal failure Renal insufficience	Kidney necrosis Renal papillary necrosis tubula necrosis	Nephrocalcinosis Renal calculi Renal stones	Anuria dysuria
Benoxaprofen		X			
Beryllium		X			
Bismuth compounds	X	X			X
Boric acid		X			X
Bucloix acid		X			
Buniodyl (bunamiodyl)	X	X	X		X
Busulfan					X
Cadmium	X	X	X		
Caffeine		X			
Capreomycin		X			
Captopril	X	X			
Carbamazepine	X	X			

Carbenicillin	X			
Carbenoxolone		X	X	
Carbitols		X		
Carbon tetrachloride	X	X	X	
Carbutamide		X		
Cefoxitin		X		
Cefuroxime		X		
Cephalexin	X	X		
Cephaloridine		X	X	X
Cephalosporins	X	X		
Cephalothin	X	X	X	X
Cephazolin		X		
Cephradine		X		
Chloroform		X		
Chloroquin		X		
Chlorpropamide	X	X		
Chlorprothixene		X		
Chlorothiazide			X	
Chlortetracycline		X		
Cholera vaccine		X		

TABLE 40 (Continued).

Drug or chemical	Nephritis Glomerulo Interstitial Pyelo-	Kidney function abnormal Kidney damage Nephropathy Nephrosclerosis Nephrosis Nephrotoxicity Renal failure Renal insufficience	Kidney necrosis Renal papillary necrosis tubula necrosis	Nephrocalcinosis Renal calculi Renal stones	Anuria dysuria
Chromium	X	X			
Cimetidine	X	X			
Cisplatin (cis-diammine-dichloroplatinum)		X			
Clofepramine		X			
Clofibrate		X			X
Clometacin		X			
Cobalt		X			
Colchicine		X			X
Colistimathate sodium		X			
Colistin sulfate	X	X			
Copper	X	X	X		
Cotrimoxazole (tri-methoprim and sulfa-methoxazole)	X	X		X	

244

Cyclophosphamide		X	X	
Cyclosporin		X		X
Cytarabine (Ara-C)		X		
Daunomycin		X		
Daunorubicin		X		
Demethylchlortetracycline		X		
Deferoxamine		X		
Dextran (low moelcular weight)		X		X
Diacetylmorphine		X		
Diatrizoate sodium		X		X
Diclofenac		X		
Diflunisal	X	X	X	
Diethylene glycol	X	X		
Diglycoaldehyde		X		
Dihydrotachysterol			X	
Dimercaprol		X		
Dioxane		X		
Diphenylhydantoin	X	X		
Dipropylene glycol		X		
Disodium edetate		X	X	

245

TABLE 40 (Continued).

Drug or chemical	Nephritis Glomerulo Interstitial Pyelo-	Kidney function abnormal Kidney damage Nephropathy Nephrosclerosis Nephrosis Nephrotoxicity Renal failure Renal insufficience	Kidney necrosis Renal papillary necrosis tubula necrosis	Nephrocalcinosis Renal calculi Renal stones	Anuria dysuria
Dithiazanine iodine		X			
Doxorubicin		X	X		
Doxycycline		X(?)			
Enalapril		X(?)			
Enflurane		X			
Ergocalciferol		X			
Erthromycin		X			
Ethambutol		X			
Ethyl acetate		X			
Ethylene dichloride	X	X			
Ethylene glycol		X	X		
Fenamic acid		X			

Substance				
Fenclofenac			X	
Fenoprofen		X	X	
Fentanyl citrate			X	
Ferbam			X	
Fibrinolysin			X	
Floctafenine			X	
Flufenamic acid	X			
Fluoride			X	
Formaldehyde	X			
Furosemide (frusemide)			X	X
Gallium nitrate			X	
Gamma globulin			X	
Gentamicin		X	X	
Glafenine			X	X
Glutathione transferase			X	
Glycerol			X	
Gold compounds	X		X	X
Guanethidine			X	
Halothane			X	
Hematin			X(?)	
Hexadimethrine		X	X	

TABLE 40 (Continued).

Drug or chemical	Nephritis Glomerulo Interstitial Pyelo-	Kidney function abnormal Kidney damage Nephropathy Nephrosclerosis Nephrosis Nephrotoxicity Renal failure Renal insufficiency	Kidney necrosis Renal papillary necrosis tubula necrosis	Nephrocalcinosis Renal calculi Renal stones	Anuria dysuria
Hydralazine	X	X			
Hydrochlorothiazide		X			
Hydroxyurea		X			
Ibuprofen		X	X		
Indandiones		X		X	
Indomethacin	X	X	X		
Iodine		X			
Iodipamide		X			
Iopanoic acid (iopadate sodium or calcium)		X			
Iron	X	X			
Iron dextran					X
Isoflurane					X

	1	2	3	4
Isoniazid			X	
Isopropyl alcohol			X	
Kanmycin		X	X	X
Ketophenbutazone			X	
Ketoprophen	X		X	
Lead compounds		X	X	X
Levamisole			X	
Lithium			X	
Magnesium trisilicate		X	X	
Manganese			X	
Mannitol			X	X
Mobendazole			X	X
Mefenamic acid			X	
Meglumine diatrizoate			X	X
Meglumine isothalamate			X	
α-Mercaptopropionylglycine			X	
Mercaptopurine			X	X
Mercurial compounds		X	X	X
Merthiolate		X		
Methanol			X	X

TABLE 40 (Continued).

Drug or chemical	Nephritis Glomerulo Interstitial Pyelo-	Kidney function abnormal Kidney damage Nephropathy Nephrosclerosis Nephrosis Nephrotoxicity Renal failure Renal insufficience	Kidney necrosis Renal papillary necrosis tubula necrosis	Nephrocalcinosis Renal calculi Renal stones	Anuria dysuria
Methicillin	X	X	X		X
Methimazole		X			
Methiodal sodium					X
Methotrexate		X			
Methoxyflurane		X	X		
Methsuximide		X			
Methyl-CCNU		X			
Methyl cellosolve	X	X			
Mehtyl chloroform		X			
Methylene blue				X	
Methyl methacrylate		X			
Methysergide		X			

Metolazone			X	
Metrizoate		X		
Metyrosine			X	
Miconazole			X	
Minocycline			X	X
Mithramycin			X	
Mitomycin			X	
Nafcillin			X	X
Naproxen			X	X
Neomycin		X	X	X
Niflumic acid			X	
Nitrofurantoin			X	X
Nitrosoureas			X	
Nomifensine			X	
Oral cholecystographic agents	X			
Oral contraceptives			X	
Osmic acid			X	
Oxacillin			X	X
Oxamniquine			X	
Oxazolidinediones			X	

TABLE 40 (Continued).

Drug or chemical	Nephritis Glomerulo	Interstitial	Pyelo-	Kidney function abnormal	Kidney damage	Nephropathy	Nephrosclerosis	Nephrosis	Nephrotoxicity	Renal failure	Renal insufficience	Kidney necrosis	Renal papillary necrosis	Renal tubula necrosis	Nephrocalcinosis	Renal calculi	Renal stones	Anuria	dysuria
Oxyphebutazone											X			X					
Oxytetracycline											X								
Paraldehyde											X								
Paramethadione											X								
Paromomycin sulfate											X								
Penicillamine		X									X								
Penicillin G		X									X								
Penicillins		X									X								
Pentamidine											X								
Pentazocine											X								
Phenacemide		X									X								
Phenazopyridine HCl											X			X					
Phencyclidine											X								

Phenidione	X	X			
Phenobarbital	X	X			
Phenolphthalein				X	X
Phenylbutazone	X	X	X		X
Phenylpropanolamine	X				
Phosphates					X
Picric acid					X
Piroxicam		X			
Polymyxin B	X	X	X		
Polymyxins		X	X	X	
Polyvinyl alcohol		X			
Potassium iodide		X			
Potassium perchlorate		X			
Povidone-iodine		X			
Practolol	X	X			
Probenecid	X	X			
Procainamide	X	X			
Procarbazine HCl	X	X			
Propylene glycol	X	X			
Propylthiouracil	X	X			
Puromycin	X	X			

TABLE 40 (Continued).

Drug or chemical	Nephritis Glomerulo Interstitial Pyelo-	Kidney function abnormal Kidney damage Nephropathy Nephrosclerosis Nephrosis Nephrotoxicity Renal failure Renal insufficience	Kidney necrosis Renal papillary necrosis tubula necrosis	Nephrocalcinosis Renal calculi Renal stones	Anuria dysuria
Pyrazinamide	X	X			
Pyrazolone derivatives		X			
Pyritinol	X				
Quinidine		X			
Quinine		X			
Rabies vaccine		X			
Radiographic contrast media (i.v. and intra-arterial injection)		X	X		
Ranitidine		X(?)			
Rifampin (rifampicin)	X	X	X		
Ristocetin		X			
Salicylates	X	X	X		
Selenium		X			

Silver compounds	X	X		
Sisomicin		X		
Smallpox vaccine	X	X		
Stibophen		X		
Streptokinase		X		
Streptomycin sulfate	X	X	X	
Streptozotocin		X	X	
Sucrose	X	X		
Sulfadiazine silver		X		
Sulfinpyrazone	X	X		
Sulfonamides (sulfadiazine, sulfamethoxazole, sulfisoxazole, etc.)	X	X	X	X
Sulindac		X		
Suramin sodium	X	X		
Tannic acid			X	
Technetium		X		
Tetrachlorethylene	X	X		
Tetracyclines	X	X	X	
Thallium	X	X		
Thiazide diuretics	X	X		

TABLE 40 (Continued).

Drug or chemical	Nephritis Glomerulo Interstitial Pyelo-	Kidney function abnormal Kidney damage Nephropathy Nephrosclerosis Nephrosis Nephrotoxicity Renal failure Renal insufficience	Kidney necrosis Renal papillary necrosis tubula necrosis	Nephrocalcinosis Renal calculi Renal stones	Anuria dysuria
Thorium dioxide	X				
Ticrynafen		X			
Tobramycin		X			
Tocainide	X				
Tolbutamide		X			
Tolmetin	X	X			
Tranexamic acid		X			
Triameterene	X			X	
Triethylenethiophos-poramide		X			
Trimethadione	X	X			
Trimethoprim (see Cotrimoxazole)	X				

	(1)	(2)	(3)	(4)
Uranium	X			
Vanadium		X		
Vancomycin	X	X	X	
Viomycin		X	X	
Vincristine		X		
Vinyl ether		X		
Vitamin A (hypervita-minosis A)		X	X	
Vitamine D (hypervita-minosis D)		X		
Vitamin K		X		
Xanthines		X		
X-ray contrast media (see Radiographic contrast media)		X	X	X
Xylitol		X		
Zomepirac		X		
Zoxazolamine	X	X		

Source: Refs. 6 and 20, and M. N. G. Dukes, and J. Elis, Eds., 1980–1984, *Side Effects of Drugs Annual*, 4–8, Excerpta Medica, Amsterdam.

(69, 70). In contrast to this, several investigators have suggested that caffeine may reduce renal function abnormalities induced by aspirin or phenacetin by virtue of its diuretic effect (71—74). Whether this discrepancy is due to different dosages used or the different natures of experiemnts conducted remains to be ascertained.

B. Antibacterial and Antifungal Agents

1. Sulfonamides

Crystalluria, hematuria, and renal failure were the most frequently observed adverse reactions associated with early sulfonamides, such as sulfadiazine, sulfathiazole, and sulfapyridine, due to their poor solubility at the urinary pH. When alkalinization of the urine became a regular part of treatment, it was soon recognized that crystalluria was not the only mechanism by which sulfa drugs cause acute renal failure (59). Both acute interstitial nephritis and tubular necrosis have been reported for various sulfonamide preparations (55, 57). It is worth noting that reported adverse drug reactions of sulfonamides include drug fever and effects on the blood, bone marrow, liver, skin, and peripheral nerves, as well as kidney (76—78). Deterioration in renal function has also been reported in patients receiving the trimethoprim-sulfamethoxazole combiantion (75).

2. Aminoglycosides

The aminoglycoside antibiotics (neomycin, gentamicin, amikacin, kanamycin, sisomycin, tobramycin, streptomycin, vancomycin) are both nephrotoxic and ototoxic. The nephrotoxic properties of these antibiotics can be attributed, at least in part, to the presence of highly water soluble sugar moieties and several basic functional groups which will be predominantly protonated, thus rendering these antibiotics as polycationic species (see Figure 30).

According to the pH-partition hypothesis (79) these poly-basic antibiotics will have relatively high urine/blood ratio. It is interesting to note that renal cortical concentrations of gentamycin are 20 times higher than plasma concentrations (80–81). In the plasma, gentamycin is estimated to be protein bound to the extent of 15 to 20%, while in the renal cortex, the extent of protein binding is as high as 70 to 85%. (82).

Combined use of another potentially nephrotoxic antibiotic such as cephalosporin with an aminoglucoside, may precipitate acute renal failure presumably by a synergistic effect, especially with preexisting renal insufficiency.

Using continuous intravenous infusion, Bodey et al. (83) have reported that daily administration of doses of at least 300 mg of gentamycin and tobramycin per square meter and 160 mg of sisomicin per square meter, nephrotoxicity occurred in only 3 out of 11 cancer patients. Eight of the 11 infections were cured.

3. *Penicillins and Cephalosporins*

The β-lactam antibiotics are relatively nontoxic and even in high doses have none or only slight nephrotoxicities in humans (84, 85). Penicillin and semisynthetic penicillins (methicillin, oxacillin, ampicillin, etc.) have in some cases been known to cause interstitial nephritis, glomerulitis, vasculitis, and renal failure, usually associated with fever, rash, and eosinophilia. There may also be accompanying proteinuria, hematuria, and pyruia. The underlying mechanism is probably allergic in nature (55, 86–90). Cross sensitization leading to an exacerba-

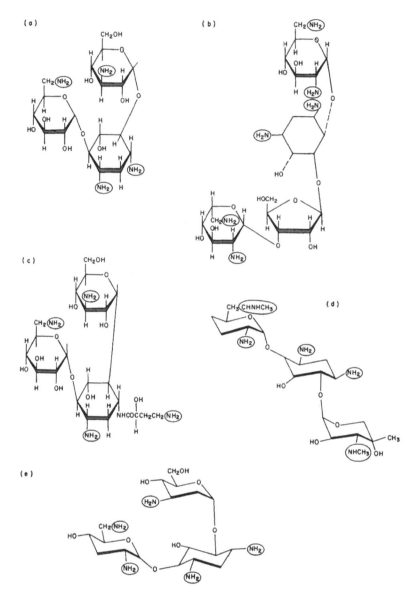

Figure 30   Structural formulas of some aminoglycoside antibiotics known to cause nephrotoxicities.   (a) Kanamycin A; (b) neomycin A; (c) amikcacin; (d) gentamicin C; (e) tobramycin.   Note the presence of several basic amino groups as well as sugar moieties.   (Adapted from Ref. 6, with permission from Blackwell Scientific Publications Ltd.)

tion of the nephropathy has been known for different penicillin
derivatives or even at times with a cephalosporin (55).

High doses of the sodium or potassium salts of penicillins
may cause electrolyte imbalance. Cardiac arrests have been
reported due to hyperkalemia following rapid intravenous injection of 20 million units of potassium penicillin G (91). To the
contrary, hypokalemia leading to fatal arrhythmias has been
induced by carbenicillin. This may be due to either urinary
excretion of potassium or through a redistribution of potassium
within body compartments (92, 93). These clearly demonstrate
the great importance of selecting the right salt of penicillin as
the dosage form and the need for careful calculation of the
critical electrolytes before an injection is given. Another
pattern of renal damage is acute renal insufficiency with complete anuria shortly following a single dose of penicillin in
patients with underlying renal disease (55). The nephrotoxic
site of antibiotics and analgesics are shown in Figure 31.

4. Tetracyclines

The tetracyclines, especially oxytetracycline, may cause pro-
gressive azotemia through their antianabolic effect on metabolism
(55, 94). Demethylchlortetracycline (now obsolete) has been
shown to cause vasopressin-resistant nephrogenic diabetic in-
sipidus. The use of degraded tetracyclines (containing 4-
epitetracycline, 4-epianhydrotetracycline, and anhydrotetracy-
cline) has been shown to cause a reversible Fanconi syndrome
in animals with glycosuria, aminoaciduria, phosphaturia, and
proteinuria, usually accompanied by tubular acidosis and occa-
sional nitrogen retention and hypokalemia. It has been suggested
(95, 96) that in prescribing tetracylines, attention should be

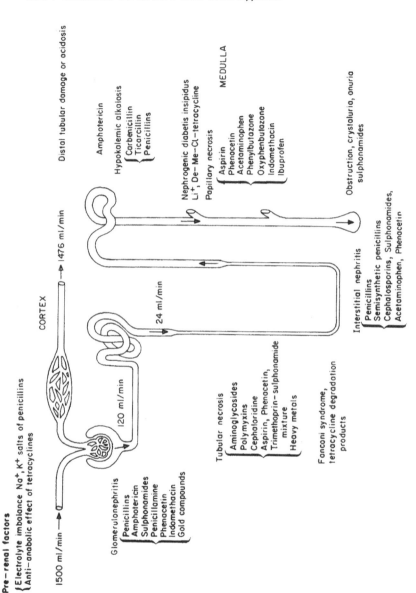

Figure 31 Nephrotoxic sites of various drugs and nephrotoxins. (Adapted from Ref. 6, with permission from Blackwell Scientific Publications Ltd.)

given by the prescriber not to exceed the amount which is
needed for one illness. One should also instruct the patient to
discard the leftover medication.

5. *Polyene Antifungal Antibiotics*

Amphotericin B is known to cause nephrotoxicity (97) as well as
hepatotoxicity, skin sensitization, and hemolysis (3, 5). Because of the lipophilic polyene moiety, amphotericin B is more
than 90% protein bound and is consequently not dialyzable and
has a relatively long half-life of 24 hr (98, 99). The nephrotoxicity appears to be dose-related, with considerable individual
variability (100—102).

6. *Polymyxins, Colistimethate, and Bacitracin*

Polymyxin B and Colistin (Polymysin E) are basic polypeptide
antibiotics (Figure 32) effective against *Pseudomonas* species
and other gram-negative bacteria (103). Both are potentially
nephrotoxic in patients with compromised renal function or if
dose exceeds 3 mg/kg/day (104). Nephrotoxicity was found to
be dose-related and was negligible at 2 mg/kg/day. Albuminuria,
azotemia, cellular casts, and loss of concentrating ability were
noted in 22 case studies by Schoenbach and co-workers (105).
Nephrotoxicity of polymyxins centers around injury to the con-
voluted tubules (106).

Colistimethate, the methanesulfonate derivative of colistin
(Figure 33), is given parenterally. Fekety et al. (107) reported
that none of the patients with good renal function developed
signs of nephrotoxicity when treated with colistimethate, and
most patients with previous renal disease tolerated the drug

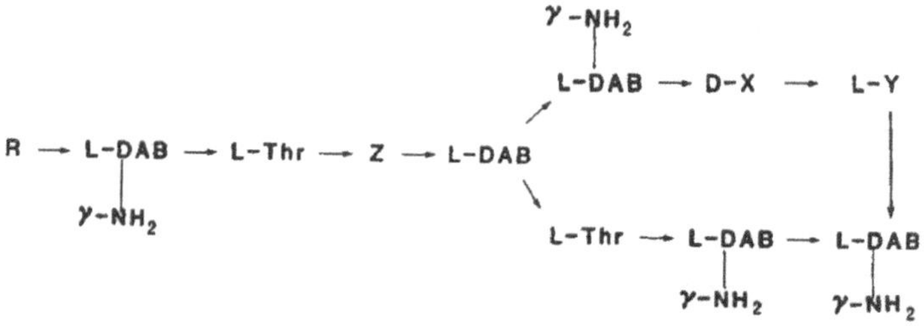

$$\gamma\text{-NH}_2$$
$$|$$
$$\text{L-DAB} \longrightarrow \text{D-X} \longrightarrow \text{L-Y}$$
$$\diagup$$
$$\text{R} \longrightarrow \text{L-DAB} \longrightarrow \text{L-Thr} \longrightarrow \text{Z} \longrightarrow \text{L-DAB}$$
$$|$$
$$\gamma\text{-NH}_2$$
$$\diagdown$$
$$\text{L-Thr} \longrightarrow \text{L-DAB} \longrightarrow \text{L-DAB}$$
$$|\qquad\qquad|$$
$$\gamma\text{-NH}_2 \qquad \gamma\text{-NH}_2$$

Polymixins

Polymyxin B_1. $C_{56}H_{98}N_{16}O_{13}$
R = (+)-6-methyloctanoyl;
X= phenylalanine   Y = leucine
Z = L-DAB.
DAB= α, γ-diaminobutyric acid

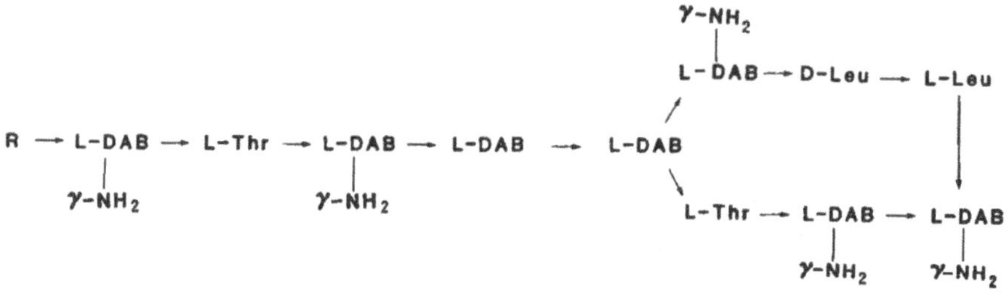

$$\gamma\text{-NH}_2$$
$$|$$
$$\text{L-DAB} \longrightarrow \text{D-Leu} \longrightarrow \text{L-Leu}$$
$$\diagup$$
$$\text{R} \longrightarrow \text{L-DAB} \longrightarrow \text{L-Thr} \longrightarrow \text{L-DAB} \longrightarrow \text{L-DAB} \longrightarrow \text{L-DAB}$$
$$|\qquad\qquad\qquad\qquad|$$
$$\gamma\text{-NH}_2 \qquad\qquad \gamma\text{-NH}_2$$
$$\diagdown$$
$$\text{L-Thr} \longrightarrow \text{L-DAB} \longrightarrow \text{L-DAB}$$
$$|\qquad\qquad|$$
$$\gamma\text{-NH}_2 \qquad \gamma\text{-NH}_2$$

Colistin

Colistin A. $C_{53}H_{100}N_{16}O_{13}$.
polymyxin E . R = (+) methyloctanoyl.

Figure 32   Molecular structures of basic polypeptide antibiotics.

quite well. However, 5 of the 24 azotemic patients under treatment developed an alarming increase in blood urea nitrogen.
The condition returned to normal when the drug was discontinued
and other causes of renal insufficiency were corrected.

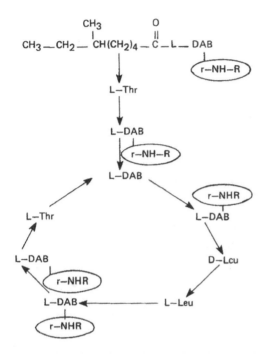

Figure 33 Chemical structure of colistimethate sodium USP, where DAB = α,γ-diaminobutyric acid and R = CH_2SO_3Na. (Note the presence of one D-amino acid and five methanesulfate groups on the γ-amino groups of DABs.)

Reduced dosages of sodium colistimethate ranging from 30 to 75% of recommended daily dose have been recommended (108) for patients with renal insufficiency (creatinine clearance from 5 ml/min or less to 20 ml/min, respectively).

From tissue binding studies, Kunin and Bugg (109) reported that polymyxin B and Colistimethate differ markedly in binding properties, probably due to the protection of the free amino groups in the latter compound (compare Figures 32 and 33).

Bacitracin is another polypeptide with an antibacterial spectrum similar to that of penicillin but resistant to penicillinase. Because of its high nephrotoxicity, it is not used intravenously (110). It is mainly used in topical application or local infiltration.

7. Vancomycin

Vancomycin was developed in response to the need for new drugs effective against staphlococcus resistant to penicillin, erythromycin, and tetracyclines. It is one of the most potent antibiotics available against certain bacteria. It is a higher molecular weight compound (MW ≈3300) with 7% nitrogen and 16 to 17% carbohydrate. With early samples of impure vancomycin, slight renal irritation and drug fever were noted in a few instances. Signs of nephrotoxicity have been much less with the newer improved commercial preparations (111, 112). Transient rise in blood urea nitrogen and severe renal damages with death due to uremia have been observed occasionally (112). Permanent deafness may occur in the presence of excessively high blood level (90 μg/ml) or higher (112). While the ototoxicity due to vancomycin has been clearly established in humans, its nephrotoxicity is still under debate (55).

C. Radiographic Contrast Media (Diatrizoate, Iodipamide, Ipodate Calcium or Sodium, Metrizoate Sodium)

These diagnostic drugs are polyiodinated carboxylic acids with both polar and nonpolar groups, used for outlining the gallbladder and bile ducts by x-ray (cholecystography and cholangiography).

Many older and more toxic contrast media have been
associated with acute renal failure and are no longer available
(113, 114). The newer preparations rarely cause acute renal
failure in the absence of predisposing factors such as diabetes
mellitus or multiple myeloma (115). Rare cases of acute renal
failure have been reported following angiography (116–118).

Intravenous pyelography is considered to be a safe procedure
by some investigators (119, 121), but acute tubular necrosis
has been reported in a patient with acute renal failure complicat-
ing pyelography intravenous infusion in a renal transplant
recipient (122). Krumlovsky et al. (115) reported 14 cases of
acute renal failure associated with radiographic contrast media,
eight within a 15-month period, although no patient had multiple
myeloma and only three were diabetic. The uricosuric effect of
these polyiodinated carboxylic acids may be responsible for the
production of renal toxicities. Iopanoic acid and ipodate calcium
have the most prominent uricosuric action (comparable to
probenecid), while iodipamide and diatrizoate sodium are less
potent (115, 123, 124). Predisposing factors and preventive
measures were suggested by these investigators (115).

V. OTOTOXICITY

It is well recognized that many drugs as well as heavy metals
and industrial solvents and chemicals may cause ototoxicities,
including deafness (7). Friedman (125) stated that it is almost
impossible to give a complete listing of chemical substances and
drugs that may cause deafness. The Boston Collaboration Drug
Surveillance Program has, on the other hand, reported the

frequency of drug-induced deafness in patients in the United States to be 3 per 1000 and 1.6 per 1000, in two reports published in 1973 and 1977, respectively. The most frequently incriminated drugs are acetylsalicylic acid (aspirin) and amino-glycoside antibiotics such as neomycin, kanamycin, and gentamicin (7). From a literature survey, over 130 (about 7.8%) drugs and chemicals have been associated with ototoxicities. The major classes are basic aminoglycoside and other antibiotics, anti-inflammatory drugs, antimalarials, β blockers, antineoplastic agents, heavy metals, diuretics, some topical agents, and various miscellaneous drugs (7).

Ototoxic chemicals and drugs may damage both vestibular or auditory (cochlear) mechanisms (see Table 41). The cochlear dysfunction is far more incapacitating. Depending on the drugs used and the mechanism of action, the inner-ear deafness is ofen preceded by symptoms such as tinnitus, diplacusis, or a feeling of fullness in the ear (126). The extent of drug-induced hearing loss depends on the dosages used, the length of exposure, as well as predisposing factors such as impaired renal function (127, 128). It is noteworthy that many ototoxic drugs are also nephrotoxic (see Table 41), but the opposite is not necessarily true. Physiologically, both the cochlea and the kidney have efficient transport systems for water and electrolytes across cell membranes. Pathologically, potassium bromate poisoning and Alport's disease damage these two organs selectively (129).

Possible mechanisms of action of drug-induced ototoxicity include inhibition of protein synthesis, the glycolytic cycle, the tricarboxylic acid (TCA) cycle, energy utilization, energy generation and the respiratory system within the mitochondria membrane of the hair cell, and alteration of the permeability of the endolymphatic membrane or alteration of the excretion system for the basic aminoglycosides in the lateral wall of the membranous cochlea. The relative rank order nephro- and ototoxicity and reactivity toward mucopolysaccharides of five aminoglycosides has been found to be related to the number of basic groups in each molecule (7) (see Figure 34).

Considering the facts that strong basic aminoglycosides do not readily penetrate the cell (130) and accumulation of drugs had been demonstrated in membrane structure (131, 132). Schacht reasoned that the initial action of these antibiotics may therefore occur at the cell membrane, altering the metabolism of membrane phosphoinositides (136) and secondarily involving membrane permeability due to phosphoinositide capacity to bind Ca^{2+} (133). Schacht proceeded to develop the dual mechanism of neomycin action, demonstrating equivalent actions for other aminoglycoside antibiotics (see Fig. 39). Normally, hydrolysis of monoester phosphate groups liberates Ca^{2+} and changes membrane permeability to cations (upper right-hand corner of Figure 35). Rephosphorylation by ATP completes the cycle (134). Neomycin occupies the calcium binding site with part of its molecule, its other basic amino groups binding to anionic sites of the lipid and possibly other membrane components. Neomycin may thus disturb membrane structure and permeability, preventing the dephosphorylation/phosphorylation cycle. It may displace calcium from the membrane and inhibit its reuptake (also gentamicin, kanamycin, tobramycin, dihydrostreptomycin, and streptomcyin). The calcium inhibition and displacement was confirmed by the

TABLE 41 Drugs That Have Been Reported to Cause Ototoxic Side Effects

Drug name	Ototoxicity	Deafness	Vestibular damage	Cochlear damage	Renal toxicity
Acetazolamide	X				X
Acetylsalicylic acid	X	X		X	X
Aconite	X				
Acinomycin C	X				
Actinomycin D	X				
Alcohol	X				
Amikacin	X				X
Aminoglycoside antibiotics	X				X
6-Aminonicotinamide	X				
Ampicillin (in hypersensitive patients)	X				X
Aniline	X				
Antimonial compounds	X				X
Apazone	X				
Arsenic compounds	X				X
Atoxyl (an arsanilic acid)	X				

Drug		
Atropine	X	
Bacitracin	X	X
Barbiturates	X	
Benorylate	X	
Benzakonium chloride	X	
Benzene	X	
Bleomycin	X	
Bonain's solution (topical)	X	
Brompheniramine	X	
Bumetanide	X	
Butikacin	X	
Caffeine	X	X
Camphor oil	X	
Capreomycin	X	X
Carbamazepine	X	X
Carbondisulfide	X	
Carbon monoxide	X	
Cisplatin	X	X
Chenopodium oil	X	

TABLE 41 (Continued).

Drug name	Ototoxicity	Deafness	Vestibular damage	Cochlear damage	Renal toxicity
Chloramphenicol	X			X	
Chlormethine	X				
Chloroform	X				X
Chloroquine	X			X	
Chlorpheniramine	X				
Chlorphentermine	X				
Chlortetracycline	X				
Clonazepam	X		X		
Colistin (polymyxin)	X		X	X	X
Co-trimoxazole (in hyper-sensitive patients)	X				X
Dideoxykanamycin B	X				
Dihydrostreptomycin	X		X	X	X
Dimethylsulfoxide	X(?)				
Digoxin	X				
Diphtheria toxoid	X				

Drug	1	2	3	4	5
Dimethyl formamide	X				
Ergot	X				
Erythromycin	X				X
Erythromycin ethylsuccinate	X				
Erythromycin lactobionate	X				
Erythromycin propionate	X				
Erythromycin stearate	X				
Ethacrynic acid	X	X	X		
Ethionamide	X				
Ethotoin	X		X		
Fluorocitrate	X				
Formaldehyde-gelatin sponge	X				
Formalin	X				
Framycetin (neomycin B)	X				X
Furosemide (in renal failure)	X				X
Gallium nitrate	X				
Gentamicin	X	X	X	X	X
Gentamicin G-1	X				
Gold salts	X				X

TABLE 41 (Continued).

Drug name	Ototoxicity	Deafness	Vestibular damage	Cochlear damage	Renal toxicity
Gramicidin	X				
Griseofulvin	X				
Hexadimethrine	X				X
Hydrocyanide	X				
Ibuprofen	X				X
Indomethacin	X				X
Insulin	X				
Iodine	X				
Iodochlorhydroquinolone	X				
Iodoform	X				
Iprindole	X				
Kanamycin	X	X	X	X	X
Kanamycin A	X				
Lead compounds	X				X
Lignocaine	X				
Lithium	X				X

Mazindol	X			
Medroxyprogesterone	X			
Mercurial compounds	X			X
Mercuric chloride	X			
Methyl mercury	X			
Minocycline	X			
Minoxidil	X			
Minsonidazole	X			
Morphine	X			
Mumps vaccine	X			
Mustine hydrochloride	X	X		
Nalidixic acid	X			
Naproxen	X			X
Neomycin	X	X	X	X
Netilmycin	X			
Nicotine	X			
Nitrobenzene	X			
Nitrous oxide	X			
Nortriptyline	X			

TABLE 41 (Continued).

Drug name	Ototoxicity	Deafness	Vestibular damage	Cochlear damage	Renal toxicity
Oral contraceptive agents	X				X
Ouabain	X				
Oxamniquine	X				
Oxytetracycline	X				X
Paromomycin	X			X	X
Perhexiline	X				
Phenazone	X				
Phenobarbital			X		X
Phenothiazines	X				
Phenylbutazone	X			X	X
Phenylephrine HCl·	X				
Phenylpropanolamine	X				X
Phenytoin			X		X
Poliomyelitis vaccine (oral)	X				
Polyethylene glycol 400	X				
Polymyxins (see Colistin)					

Potassium bromate	X			
Practolol	X			X
Procaine	X			
Propranolol	X			
Propylene glycol	X			
Propylthiouracil	X			X
Quinidine	X			X
Quinine	X			X
Ribostamicin	X			
Rifampicin	X			X
Ristocetin	X			X
Sagamicin	X			
Salicylates	X			X
Salvarsan	X			X
Scopolamine	X			
Sisomicin	X			X
Streptomycin	X	X	X	X
Strychnine	X			
Tetanus toxoid	X			

TABLE 41 (Continued).

Drug name	Ototoxicity	Deafness	Vestibular damage	Cochlear damage	Renal toxicity
Tetracycline hydrochloride	X				X
Thalidomide	X				
Tobramycin	X				X
Tolmetin	X				X
Tranexamic acid	X				X
Tricyclic antidepressants	X				
Valerian	X				
Vancomycin	X			X	X
Viomycin (highly ototoxic)	X		X	X	X
Vitamin A	X				X

Source: Refs. 7 and 41 and M. N. G. Dukes and J. Elis, Eds., 1980—1984, *Side Effects of Drugs Annual 4—8*, Excerpta Medica, Amsterdam.

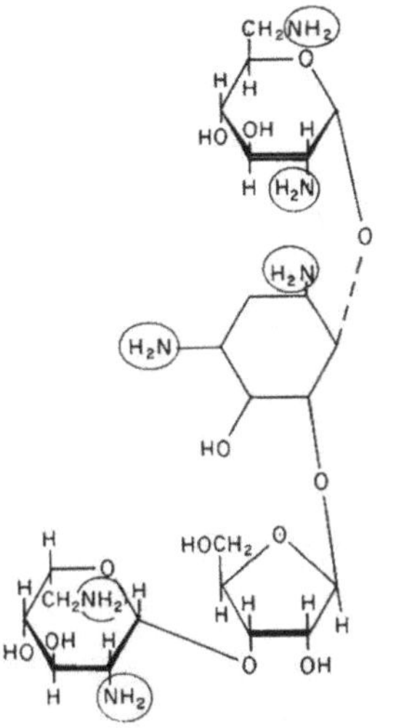

Neomycin, with 6 basic amino groups

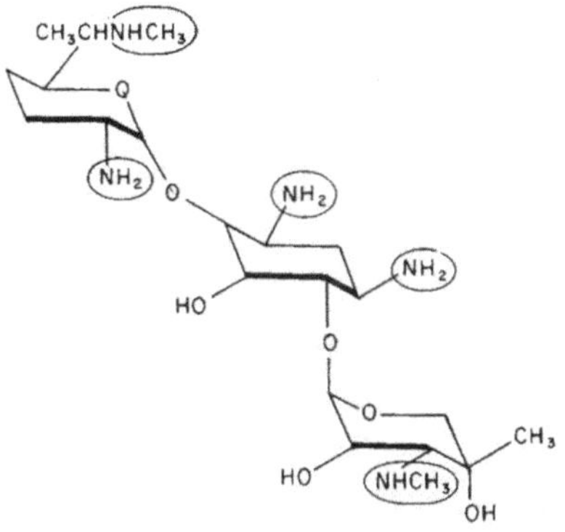

Gentamicin, with 5 basic amino groups

Sagamicin, $C_{20}H_{41}N_5O_7$ similar to gentamicin

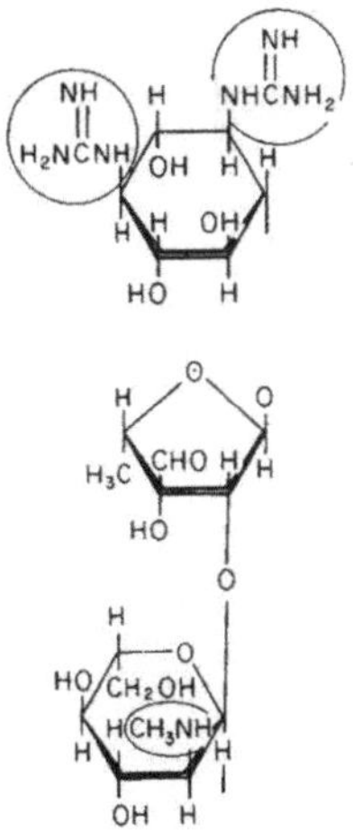

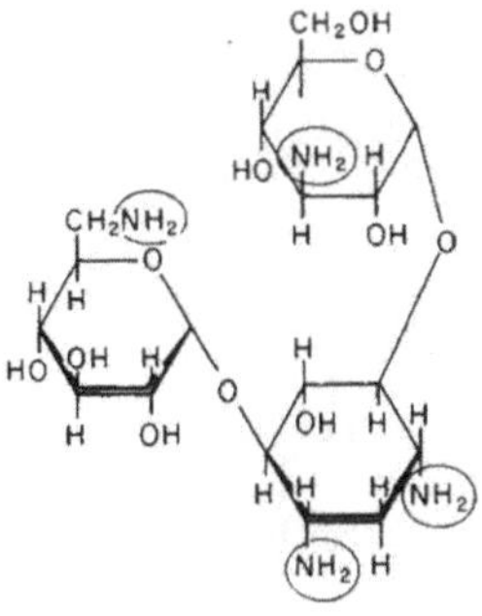

Kanamycin, with 4 basic amino groups

Streptomycin, with 2 guanidino groups and 1 amino group

Figure 34 Oxotocic aminoglycosides with different number of basic groups per molecule.

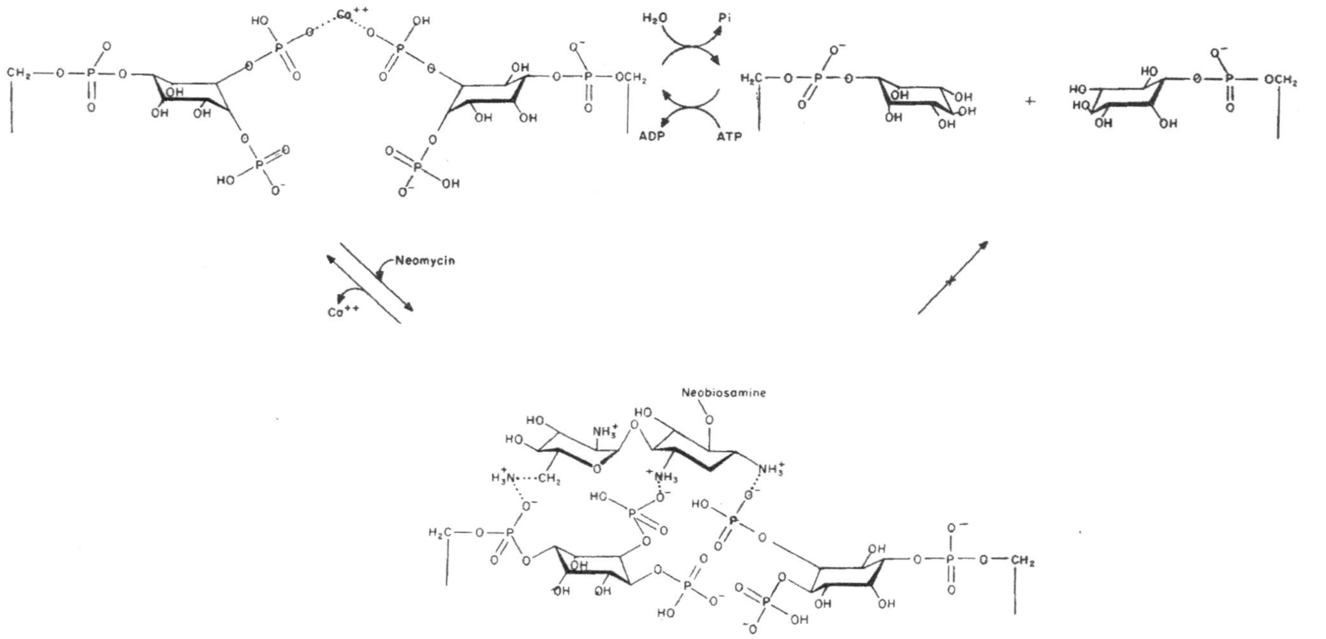

Figure 35 Mechanism of action of neomycin on membrane phosphoinositides with Ca²⁺. Note the formation of three ionic bonds between the —NH₃⁺ groups of neomycin and the phosphate groups of phosphoinositide. (Adapted from Ref. 7, with permission from Blackwell Scientific Publications Ltd.)

work of Orsulakova et al. (135). With increasing drug concentration, multiple sites of drug binding may impose conformational changes on the membrane disrupting its structure and
function. Such actions on the membrane may create ionic im-
balances sufficient to cause cellular destruction. Alternatively,
the disturbance of membrane structure and permeability may
facilitate the entry of neomycin or other basic antibiotics into
the cell, where it could exert a similar or different action on
intracellular structures such as mitochondria (136).

VI. SKIN SENSITIZATION AND DRUG-INDUCED
EOSINOPHILIA

Many chemicals and drugs are known to cause dermatitis,
urticaria, eczema, photosensitization, or other signs of skin
sensitization following systemic or topical applications (see Table
42). The role of bioactivation followed by covalent bond
formation between a drug molecule and a macromolecule has been
discussed previously (8). Many drugs known to cause blood
dyscrasias are also known to cause skin sensitization (see Table
38 and reference 4).

From Table 42 one can see that many drugs capable of
causing eosinophilia are also known to cause skin sensitization
and/or lupus erythematous. Some of them are also hepatotoxic
(5) or nephrotoxic (6). Although minor skin disorders as such
may not be a clincally serious problem, drug-induced skin
problems should not be neglected. They may precede or signal
more serious adverse reactions involving other vital organs and
tissues (e.g., bone marrow, kidney, and liver) in sensitive
individuals.

TABLE 42 Drugs Known to Cause Skin Sensitization and/or Eosinophilia

Drug	Skin sensitization	Lupus erythematosus	Eosinophilia
Allopurinol	X		X
Amidopyrine	X		X
Aminoglutethimide		X	
Aminophenazone			X
Aminosalicylic acid		X	X
Amobarbital			X
Amoxicillin	X		X
Amphotericin B			X
Ampicillin	X	X	X
Aniline			X
Antazoline			X
Arestatin	X		
Arsenicals, inorganic			X
Arsenicals, organic			X
Aspirin			X
Atenolol	X		
Azathioprine			X
BCG vaccine	X		
Benoxaprofen	X		
Benzene			X
Benzindazole	X		
Benzonatate	X		
Benzoyl peroxide	X		
Benztropine	X		
Brompheniramine	X		
Butaperazine	X		X

TABLE 42 (Continued).

Drug	Skin sensitization	Lupus erythematosus	Eosinophilia
Captopril	X		
Carbamazepine	X		X
Carbenicillin			X
Carbimazole			X
Carphenazine	X		
Cefamandole	X		
Cefazolin	X		X
Cephalexin	X		X
Cephaloglycin			X
Cephaloridine	X		X
Cephalosporins	X		X
Cephalothin	X		X
Cephazolin	X		X
Chloramphenicol	X		
Chlordiazepoxide	X		X
Chlormezanone	X		
Chlorobutanol	X		
Chlorophenothane			X
Chloropromazine	X	X	X
Chlorothiazide			X
Chlorphenesin	X		
Chlorphentermine	X		
Chlorpropamide	X		X
Chlorprothixene	X	X	
Cimetidine	X		
Clofazimine	X		

TABLE 42 (Continued).

Drug	Skin sensitization	Lupus erythematosus	Eosinophilia
Cloxacillin	X		X
Codeine	X		
Co-trimoxazole	X		
Cromolyn sodium		X	X
Cyanocobalamin	X		
Cyclophosphamide	X		
Cyproheptadine	X		
Dapsone	X		X
Desipramine			X
Dextran			X
Diatrizoate			X
Diazoxide			X
Diclofenac	X		
Dicloxacillin	X		X
Diethylcarbamazine	X		
Diethylpropion	X		
Diflunisol	X		
Digitalis leaf			X
Dimethyl sulfoxide	X		
Diphemanil	X		
Diphenydramine	X	X	
Diphenylhydantoin			X
Dipyron	X		
Doxepin	X		
Doxycycline			X
Erythromycin			X

TABLE 42 (Continued).

Drug	Skin sensitization	Lupus erythematosus	Eosinophilia
Ethambutol	X		X
Ethopropazine	X		
Ethosuximide		X	X
Ethotoin	X		X
Etretinate	X		
Fenbufen	X		
Fenclofenac	X		
Fluorocytosine			X
Fluorouracil	X		
Fluphenazine			X
Furazolidone			X
Furosemide	X		X
Glucagon	X		
Glucocorticoids			X
Glue			X
Glutethimide	X	X	
Glycopyrrolate	X		
Glyeryl trinitrate	X		
Gold compounds			X
Griseofulvin	X	X	X
Guanidine			X
Haloperidol			X
Heroin			X
Hexachlorophene	X		
Hexachorocyclohexane			X
Hydralazine		X	X

TABLE 42 (Continued).

Drug	Skin sensitization	Lupus erythematosus	Eosinophilia
Hydroxychloroquine			X
Ibuprofen	X	X	
Imipramine			X
Indanedione derivatives	X		X
Indomethacin	X		X
Insulin		X	
Iodides	X		X
Isoniazid		X	X
Isoxsuprine	X		
Ketoconozole	X		
Levamisole	X		
Levodopa		X	X
Lithium carbonate	X		X
Mepazine			X
Mephobarbital		X	
Meprobamate			X
Mercury compounds			X
Mesoridazine	X		X
Methadone	X		X
Methaqualone	X		
Methenytoin		X	X
Methicillin	X		X
Methimazole		X	X
Methotrimeperazine		X	
Methoxyflurane			X
Methsuximide		X	X

TABLE 42 (Continued).

Drug	Skin sensitization	Lupus erythematosus	Eosinophilia
Methyldopa		X	X
Methylprednisolone			X
Methylthiouracil		X	X
Methysergide		X	
Metoprolol	X		
Metronidazole			X
Mianserin	X		
Miconazole	X		
Minoxidil	X		
Misonidazole	X		
Musk ambrette	X		
Nafcillin	X		X
Nalidixic acid	X		X
Nitrofurans			X
Nitrofurantoin		X	X
Novobiocin			X
Nystatin	X		
Oleandomycin			X
Oxacillin	X		X
Oxaprozin	X		
Oxazepam			X
Oxyphenbutazone	X	X	
Oxyprenolol	X		
Oxytetracycline		X	
Papaverine			X
Paramethadione		X	

TABLE 42 (Continued).

Drug	Skin sensitization	Lupus erythematosus	Eosinophilia
Penicillamine	X	X	X
Penicillins	X	X	X
Pentazocine	X		X
Pentobarbital			X
Perazine		X	
Perphenazine		X	X
Phenacetin		X	X
Phenindione	X		X
Phenobarbital			X
Phenolphthalein	X		
Phenothiazines	X		X
Phenylbutazone	X	X	X
Piroxicam	X		
Polymixin B			X
Polyvinyl chloride	X		
Potassium iodide	X		
Potassium perchlorate			X
Povidone iodine	X		
Practolol		X	
Primidone		X	X
Procainamide		X	X
Procarbazine	X		
Prochlorperazine			X
Promethazine		X	
Propranolol	X		X
Propylthiouracil		X	X
Protreptyline	X		

TABLE 42 (Continued).

Drug	Skin sensitization	Lupus erythematosus	Eosinophilia
PUVA therapy (psoralen)	X		
Pyrazinamide	X		
Pyritinol	X		
Quinacrine			X
Quinidine	X	X	X
Quinine	X		X
Reserpine		X	
Rifampin			X
Ristocetin			X
13-*cis*-Retinoic acid	X		
Retinol analogs	X		
Salicylazosulfa-pyridine (sulfasalazine)	X	X	X
Sodium salicylate	X		X
Spironolactone	X		X
Stibophen			X
Streptomycin	X	X	X
Sulfadiazine	X		X
Sulfadimethoxine	X		X
Sulfamerazine	X	X	X
Sulfamethizole	X		X
Sulfamethoxazole	X		X
Sulfamethoxydiazine	X		X
Sulfamethoxy-pyridazine	X		X

TABLE 42 (Continued).

Drug	Skin sensitization	Lupus erythematosus	Eosinophilia
Sulfanilamide	X		X
Sulfapyridine	X		X
Sulfasolidine	X	X	
Sulfathiazole	X		X
Sulfinpyrazone	X		
Sulfisoxazole	X	X	X
Sulfonamides	X	X	X
Sulfones	X		
Sulindac	X		
Suramin	X		
Tetracyclines	X	X	X
Thalium			X
Thiacetazone	X		
Thiamazole		X	
Thiazide diuretics	X		
Thioridazine		X	X
Thiothixene	X		X
Thiouracil		X	
Tolbutamide			X
Toluene			X
Triacetyloleando-mycin			X
Trichloroethylene			X
Trifluoperazine		X	X
Triflupromazine	X		X
Trimethadione		X	X

TABLE 42 (Continued).

Drug	Skin sensitization	Lupus erythematosus	Eosinophilia
Trimethoprim			X
Trimethoprim-sul-famethoxazole			X
Trinitrotoluene			X
Troxidone	X	X	
Venopyronum		X	
Vitamin A			X
Vitamin D			X
Vitamin E	X		
Vitamin K	X		X
Wax crayons (paranitraniline)			

Source: Refs. 5, 8, 21, and 41 and M. N. G. Dukes and J. Elis, Eds., 1980–1984, *Side Effects of Drugs Annual 4–8*, Excerpta Medica, Amsterdam.

VII. GLAUCOMA AND CATARACTS

The eye, being one of the most delicate sensory organs, is nourished by a rich supply of blood from the ophthalmic artery and its many branches. It is also controlled by an intricate balance of sympathetic and parasympathetic nervous tones. The eye is responsive to influences from both systemic and topical applications of drugs. At the time it is subject to external insults such as light, radiation, pollutants, and extreme changes in temperature and humidity (9).

Industrial chemicals (e.g., naphthalene) have been known
to cause lenticular and corneal opacities since the turn of the
century (137). During the 1930s, when dinitrophenol was used
in weight reduction, many patients developed cataracts before
the product was removed from the market (138—145).

Increased intraocular pressure or glaucoma can result from
any drug or chemical that can affect aqueous humor secretion
or outflow through the trabecular meshwork and Schlemm's
canal. Table 43 lists many different drugs and chemical reported to cause either glaucoma and/or cataracts in human subjects or experimental animals (9, 41, 146).

A. Glaucoma

Mydriatics such as atropine, belladonna, and anticholinergics,
and sympathomimetic amines such as phenylophrine and ephin-
ephrine may cause a predisposed eye with a shallow or narrow
angle to abruptly shut off the drainage of aqueous fluid following the contraction of the peripheral iris and dilation of the
pupil. Following the use of these drugs, a rapid increase in
intraocular pressure may result in acute angle-closure glaucoma
due to continued secretion of aqueous fluid in predisposed eyes.
Phenothiazines and tricyclic antidepressants may also precipitate
angle-closure glaucoma in some patients due to their anticholinergic effect. Topical corticosteroids such as 1% dexamethasone
may produce open-angle glaucoma due to increase in outflow resistance attributed to the hydration of mucopolysaccharides in
the trabecular meshwork (147, 148). While psychotropic drug
may aggravate glaucoma, with proper safeguards, these drugs
can still be used in patients with diagnosed glaucomas (149).

TABLE 43 Drugs and Chemicals Associated with Glaucoma and/or Cataracts

Drug or chemical	Glaucoma increased ocular pressure	Cataracts less deposit or opacity
Acetophenazine		X
ACTH (corticotropin)		X
Adrenal corticosteroids	X	X
Adrenergic agents	X	
Aldosterone	X	X
Allopurinol		X(?)
Alloxan		X
Aluminum nicotinate	X(?)	
Amiodarone		X
Amitriptyline	X	
Amodiaquine		X(?)
Amphetamine	X	
Amyl nitrate	X	
Anticholinergic agents (systemic)	X	
Anticholinesterase (eye drops)		X
Antihypertensives	X	
Atropine	X	
Aurothioglucose		X
Aurothioglycine		X
Belladonna	X	
Benzhexol	X	
Benztropine	X	
Betamethasone	X	X
Biperiden	X	
Dis(phenylisopropyl)-piperazine		X

TABLE 43 (Continued).

Drug or chemical	Glaucoma increased ocular pressure	Cataracts less deposit or opacity
Busulfan		X
Butaperazine		X
Butyrophenones	X	
Capreomycin		X(?)
Carbamazepine		X(?)
Carbon dioxide	X	
Carbromal		X(?)
Carbutamide		X
Carphenazine		X
Chlorobenzene		X
Chlorophenylalanine		X
Chloropropamide		X
Chloroquine		X(?)
Chlorpromazine	X(?)	X
Chlorprothixene		X
Cholesterol lowering agent		X
α-Chymotrypsin	X	
Clidinium + chlordiazepoxide	X	
Clomiphene		X(?)
Cobalt chloride		X
Colchicine		X(?)
Colloidal silver		X
Corticosteroids (topical and systemic)	X	X
Cortisone		X
Cresol		X

TABLE 43 (Continued).

Drug or chemical	Glaucoma increased ocular pressure	Cataracts less deposit or opacity
Cyclopentolate	X	X
Cyclophosphamide		X
Cyproheptadine	X	
Decahydronaphthalene		X
Deferoxamine		X(?)
Demecarium bromide	X	X
Deoxycorticosterone	X	X
Desipramine	X	
Dexamethasone	X	
Diazepam	X	X
Diazoxide		X
Dibenzazepine derivatives	X	
Dibromomannitol		X(?)
Dichlorisone		X
Dichloronitroaniline		X
Diethazine		X
3β-(β-Diethylaminoethoxy)-endrost-5-en-17-one-methoxime		X
Diethylcarbamazine		X
Diisopropyl fluorophosphate		X
4-(p-Dimethylaminostyryl)-quinoline		X
Dimethylnitroquinolone		X
Dimethylsulfoxide		X
Dimethylterephathalate		X
Dinitro-o-cresol		X

TABLE 43 (Continued).

Drug or chemical	Glaucoma increased ocular pressure	Cataracts less deposit or opacity
Dinitrophenol		X
Diquat		X
Dithizone		X
Doxepin	X	
Droperidol		X(?)
Dyflos		X
Echothiophate iodide (phosphaline iodide)		X
Edrophonium		X
Epinephrine	X	X
Ergot	X	X(?)
Erythrityl tetranitrate	X(?)	
Estrogen	X	
Ethopropazine		X
Ethotoin		X(?)
Fluorocortisone	X	X
Fluorometholone	X	X
Fluphenazine		X
Fluprednisolone	X	X
Galactose (in galactosemia)		X
Ganglionic blocking agents	X	
Glucocorticoids	X	X
Glucose		X
Glyceryl trinitrate	X	
Gold (^{198}Au) compounds		X
Gold sodium thiomalate		X

TABLE 43 (Continued).

Drug or chemical	Glaucoma increased ocular pressure	Cataracts less deposit or opacity
Haloperidol		X
Hematoporphyrin		X
Hexamethonium	X	
Homatropine	X	
Hydrocortisone	X	X
Hydroflumethiazide	X	
Hydroxyamphetamine	X	
Hydroxychloroquine		X(?)
Hyoscyamine	X	
Ibuprofen		X(?)
Imipramine	X	X
Insulin	X	
Iodoacetic acid		X
Ipratropium	X	
Isoflurophate (DFP)	X	X
Isosorbide dinitrate	X(?)	
Ketamine	X	
Lead compounds		X(?)
Levomepromazine		X
Levorphanol		X
Mannitol hexanitrate	X(?)	
Medrysone	X	X
Menthol		X(?)
Meperidine		X
Mephenytoin		X(?)

TABLE 43 (Continued).

Drug or chemical	Glaucoma increased ocular pressure	Cataracts less deposit or opacity
Mercuric oxide		X
Mesoridazine		X
Methadone		X
Methdilazine		X
Methotrexate		X
Methotrimeperazine		X
Methoxalen + UV light		X
Methyl dichlorisone		X(?)
Methylphenidate	X(?)	
Methylprednisolone	X	X
Mild silver protein		X
Mimosine		X
Miotics		X
Mitomycin	X	
Mitotane		X
Morphine	X(?)	X
Mydriatics (topical)	X	
Myleran		X
Nalidixic acid	X	
Naphthalene		X
β-Naphthol		X(?)
1,2-Naphthoquinone		X
Neostigmine		X
Neuroleptic agents	X	
Niacinamide	X(?)	

TABLE 43 (Continued).

Drug or chemical	Glaucoma increased ocular pressure	Cataracts less deposit or opacity
Nicotinic acid	X(?)	
Nicotinyl alcohol	X(?)	
Nitrates	X	
Nitrogen mustard		X
Nitroglycerin	X	
Nitrous oxide	X	
Opiates		X
Oral contraceptives	X(?)	X(?)
Oral hypoglycemic drugs		X
Orphenadrine	X	
Orphenamide	X	
Oxyphenonium	X	
Pantocaine		X
Para-dichlorobenzene		X
Paramethasone	X	X
Paraoxon		X
Parasympathomimetic drugs	X	
Penicillamine		X(?)
Pentaerythritol tetranitrate	X(?)	
Perazine		X
Pericyzaine		X
Perphenazine		X
Phenelzine + serotonin		X
Phenmetrazine		X(?)
Phenothiazines		X

TABLE 43 (Continued).

Drug or chemical	Glaucoma increased ocular pressure	Cataracts less deposit or opacity
Phentermine		X(?)
Phenylbutazone		X
3-(2-Phenyl)-hydrazopropionitrile		X
Phenylephrine	X	
Phenylmercuric acetate		X
Phenylmercuric nitrate		X
2-(4-Phenyl-1-piperazinylmethyl)-cyclohexanone		X
Physostigmine		X
Pilocarpine	X[a]	X(?)
Piperacetazine		X
Piperazine		X(?)
Polymyxin B		X
Prednisolone	X	X
Prednisone	X	X
Prochlorperazine		X
Procyclidine	X	
Promazine		X
Promethazine		X
Propantheline	X	
Propiomazine		X
Prostaglandin E_1	X	
Protriptyline	X	
Psoralens		X
Pyrithione		X

TABLE 43 (Continued).

Drug or chemical	Glaucoma increased ocular pressure	Cataracts less deposit or opacity
Radium		X
Scopolamine	X	
Silicone		X
Silver nitrate		X
Silver protein		X
Sodium chloride	X	
Streptomycin		X
Streptozotocin		X
Succinylcholine	X	
Sulfaethoxypyridazine		X
Sulfanilamide		X
Suxamethonium	X	
Sympathomimetic amines	X	
Testosterone	X(?)	
Tetraline		X
β-Tetralol		X
Thallium		X
Thiethylperazine		X
Thiopropazate		X
Thioproperazine		X
Thioridazine		X
Thiotepa		X
Thiothixene		X
Tolazoline	X	
Tolbutamide		X

TABLE 43 (Continued).

Drug or chemical	Glaucoma increased ocular pressure	Cataracts less deposit or opacity
Tretamine		X
Triamcinolone	X	
Triaziquone		X
Tricyclic antidepressants	X	
Triethylene melamine		X
Trifluoperazine		X
Trifluperidol		X(?)
Triflupromazine		X
Trihexyphenidyl	X	
Trimeprazine		X
Trinitrotoluene		X(?)
Triparanol		X
Triperidol		X
Trolnitrate	X(?)	
Tropicamide	X	
Urokinase	X	
Vitamin A	X	
Vitamin D		X(?)
Vitamin D_2		X(?)
Vitamin D_3		X(?)
Xylose		X(?)

[a]Initial increase followed by decrease in ocular pressure.
Source: Refs. 9 and 137—168 and Dukes, M. N. G., and Elis, J., Eds. (1980—1984), *Side Effects of Drugs Annual 4—8*. Excerpta Medica, Amsterdam.

Other drugs with questionable roles in producing glaucoma include nitrates, niacinamide, nicotinic acid, nicotinyl alcohol, testosterone, tropicamide, urokinase, and vitamin A.

B. Cataracts—Lens Deposit or Opacity

Many organic compounds have been shown to cause cataracts either in humans or in experimental animals. The list includes naphthalene, β-tetralol, tetraline, β-naphthol, chloro- and *p*-dichlorobenzene, dinitrophenol, and cresol (146, 150). Even natural products such as mimosine have been shown to cause cataracts in rats (151).

Inorganic compounds such as thallium, cobalt, gold, and silver preparations can also cause cataracts in rats and rabbits. This may be due to nonspecific cytotoxicities of these heavy metals, or, in the case of silver nitrate and silver protein, to photosensitivity. Since the eye is exposed to light, photo-sensitive drugs, if present in this organ, may have the potential of contributing to toxic effects, although this still remains to be proved (147). Some examples in this category are sulfonamides, carbamazepine, tricyclic antidepressants, psoralens, estrogens and progesterones, tranquilizers (chlorprothixen, haloperidol, thiothixene), and phenothiazines (chlorpromazine, fluphenazine, perphenazine, prochlorperazine, thioridazine, trifluoperazine, and triflupromazine) (152, 153). Patients taking these drugs should protect their eyes from exposure to sunshine or ultraviolet light.

Miotic agents, especially of the anticholinesterase group which are used for the treatment of glaucoma, are known to

cause lenticular opacities (146, 154, 155). These include di-
isopropyl fluorophosphate (DFP), echothiophate, demecarium
bromide, and diethyl-p-nitrophenyl phosphate (paraoxon). Although long-term use of pilocarpine has been associated with
cataracts (155, 156), the causal role of pilocarpine in cataract-
ogenicity has been questioned by medical experts, since many
patients have retained clear lenses after using this drug for
several decades (147).

Corticosteroids used systemically for the treatment of
rheumatoid arthritis and prolonged topical use have been shown
to cause nonreversible posterior subcapsular cataracts (156,
157–159). The lens opacity appears to be dependent on both
the dosage and the duration of treatment.

Epinephrine-induced opacities have been demonstrated in
experimental animals such as mice and rats. Narcotic analgesics,
such as morphine, levorphanol, meperidine, and methadone,
have been reported to cause lens opacities in rodents (160,
161). Weinstock and Scott have reported that in rodents,
opacities caused by the acute administration of narcotic analgesics,
epinephrine, chlorpromazine, and cataracts induced by oxygen
deprivation occur in the anterior portion of the lens below the
capsule. The cataractogenic effect of morphinelike drugs (but
not epinephrine or chlorpromazine) can be abolished by oxygen
and a respiratory stimulant (amiphenazole). Reserpine pretreat-
ment abolished only those opacities produced by analgesics, but
not by chlorpromazine. This suggests the involvement of
catecholamines in the mediation of opacities by narcotic analgesics.
These authors further reported that a temporary rise in the

tonicity of the aqueous humor may occur, resulting from increased evaporation of water through the cornea when sympathetic activity is stimulated and the blink rate of the eye is
reduced (162). On the other hand, the acute lens opacity induced by chlorpromazine in mice does not appear to be caused
by lid retraction but by hypothermia (163).

Several ophthalmological problems associated with the use
of oral contraceptives (glaucoma, cataracts, pseudotumor cerebri,
and retinal vein obstruction) have been reported in retrospec-
tive studies (164, 165). On the other hand, no significant
increase of ocular pathology can be found in oral contraceptive
users compared to nonusers in randomized samples in prospec-
tive studies (164, 165).

Triparanol, a drug marketed in the 1950s to lower blood
cholesterol in patients, was found to cause posterior subcapsular
lens opacities. It was withdrawn from the market in 1962 because of ocular and dermatological side effects (146, 147).

Anticancer agents such as busulfan, methotrexate, and
nitrogen mustards have also been known to cause cataracts (146,
147). Miscellaneous drugs with questionable roles in causing
cataracts or lens opacities are amodiaquine, carbamazepine,
carbromal, chloroquine, clomiphene, colchicine, diazepam, ergot,
ethotoin, hydroxychloroquine, mephenytoin, penicillamine,
piperazine, trifluperidol, and vitamins A and D.

Sugars such as galactose, glucose, and xylose are known
to be cataractogenic (166, 167). Varma et al. have further reported that inhibitors of lens aldose reductase, which catalyzes
the formation of sugar alcohol, can be used in preventing the

onset of diabetic or galactosemic cataracts (167, 168). Several flavonoids have been shown to be effective inhibitors of aldose reductase; among these, quercetin, quercitrin, and myricitrin are the most potent inhibitors of this target enzyme (169). This enzyme should serve as a viable biochemical target for the development of effective drugs for the prevention of cataracts in diabetic patients and others showing signs of developing cataracts.

VIII. CARCINOGENICITY OF ANTICANCER DRUGS

The first experimental success in treating a mouse ascites tumor (6C3HED) with nitrogen mustard, and the subsequent clinical trial on Hodgkin's disease, opened up a new vista for cancer chemotherapy (170, 171). In 1956—1961 Li, Hertz, and co-workers reported the first firm evidence that chemotherapy with methotrexate could actually cure cancers such as choriocarcinoma and related trophoblastic tumors (172, 173). Since then over a dozen different malignancies have been documented to be curable by combination chemotherapy, together with other modalities, such as surgery and radiotherapy (174).

Since many chemotherapeutic agents are mutagenic, they are also potential carcinogens. Alkylating agents are especially well known to cause second neoplasms such as nonlymphocytic leukemia and non-Hodgkin's lymphoma (175). Cyclophosphamide has also been reported to cause carcinoma of the bladder (175). From our recent literature survey (11), we have found 37 anticancer agents with reported carcinogenicity or co-carcinogenicity.

These drugs include 16 alkylating agents and eight antimetabolites, the rest being hormones, alkaloids, and miscellaneous drugs (Table 44). In some patients even tertiary cancers have been observed after chemotherapy with procarbazine. More active research in the field of preventive drugs against carcinogenesis as well as noncarcinogenic anticancer drugs is desperately needed.

Future development of more effective anticancer agents against specific types of cancer will depend on new knowledge generated in molecular biology, especially in the area of gene expression and control (176), as well as known methods based on biochemical mechanisms and enzymology (177). New multivariate statistics (178) and computer graphic techniques (179–182) will undoubtedly be useful tools in future drug design.

TABLE 44   Classification, Structure and Carcinogenic Activities of Anticancer Agents

Name (Classification)	Structure	Carcinogenic activities
Busulfan (Alkylating agent)	$H_3CO_2SO(CH_2)_4OSO_2CH_3$	Thymic lymphoma and ovarian tumor in RF mice Local sarcomas in rats
Chlorambucil (Alkylating agent)	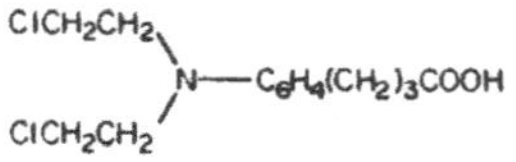 $ClCH_2CH_2$—N—$C_6H_4(CH_2)_3COOH$ / $ClCH_2CH_2$	Papillomas Lymphoma, regenerating liver in mice Lung tumors, lymphosarcomas and ovarian tumors in mice
Cyclophosphamide (Alkylating agent) Used in children with Ewing's sarcoma (9–11)	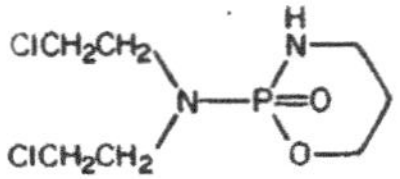	Plasmacytoma in hamster Lung tumors in mice Carcinomas, pulmonary adenomas and lymphomas in mice
Dibromomannitol (Alkylating agent)	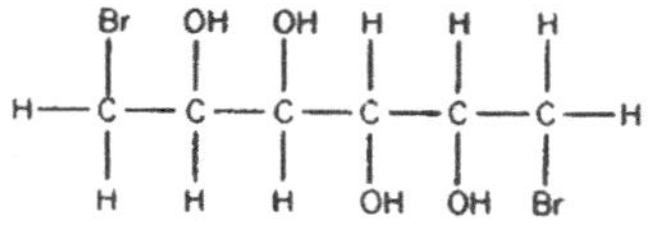	Peritoneal sarcomas, subcutaneous tumors in rats and lung tumors, lymphomas in mice
Dibromodulcitol (Alkylating agent)	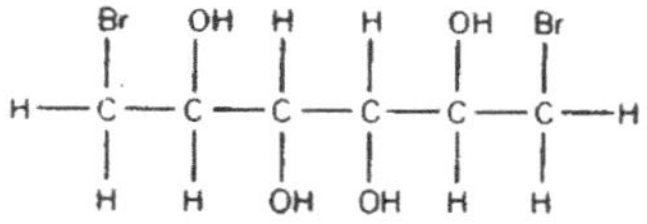	Lung tumors and lymphomas in mice and subcutaneous tumors in rats
DIC (Alkylating agent)	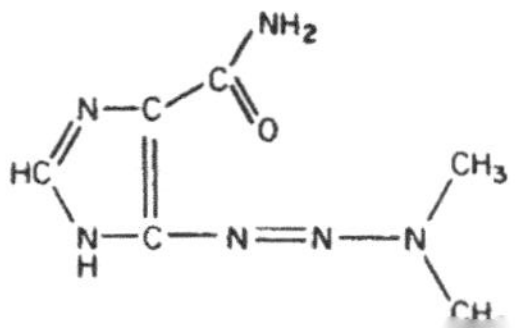	An *in vivo* methylating agent in liver, kidney, lung and brain of rats Lung tumors, lymphomas, splenic tumors and uterine tumors in mice, and lymphomas, renal tumors, heart tumors and breast carcinomas in rats

N-Hydroxyurethan
(Alkylating agent)

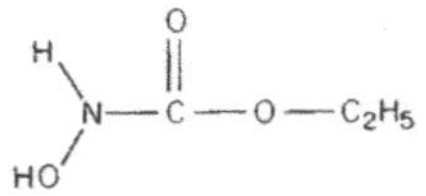

Lung adenomas, skin tumors, lung tumors, and leukaemia in mice

Isophosphamide
(Alkylating agent)

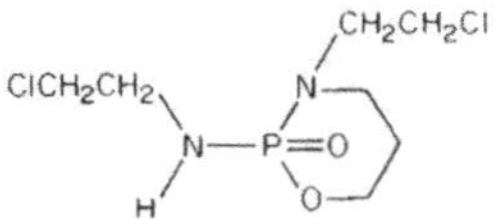

Malignant lymphomas of the haematopoietic system in mice and leiomyosarcomas of the uterus in rats

Melphalan
(Alkylating agent)

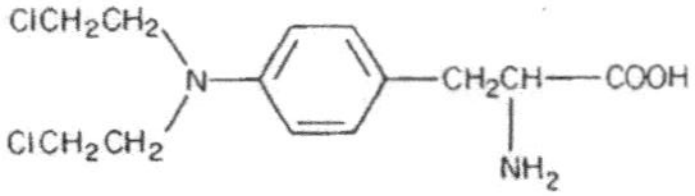

Papillomas in mouse skin
Lung tumors and lymphosarcomas in mice and peritoneal sarcomas in rats

Streptozotocin
(Alkylating agent)
(Antibiotic)

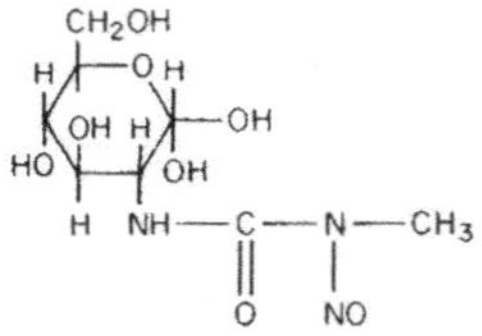

Kidney tumors, and damage of the pancreatic islet cell in rats
Biliary hyperplasia, cholangiomas and line tumours in hamster

Triethylene melamine
(Alkylating agent)

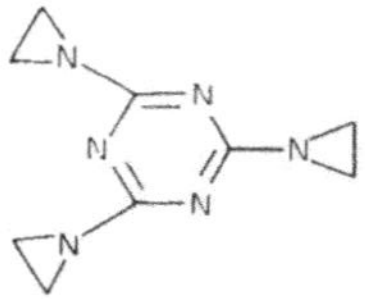

Dominant lethal mutations in male mammals
Lymphoma in mice

Uracil mustard
(Alkylating agent)

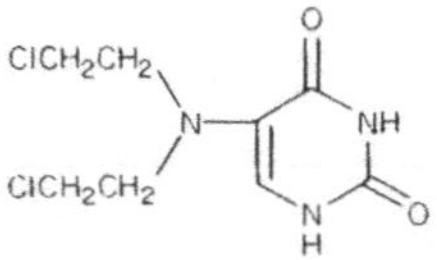

A direct-acting carcinogen

309

TABLE 44 (Continued).

Name (Classification)	Structure	Carcinogenic activities
Urethan (Alkylating agent)	$\begin{array}{c} O \\ \parallel \\ H-N-C-O-C_2H_5 \\ \| \\ H \end{array}$	A multipotent carcinogen and a 'promoting' agent Lung adenomas, skin papilomas, malignant lymphomas, hepatomas, liver haemangiomas, mammary carcinomas
Nitrogen mustard (Alkylating agent)	$\begin{array}{c} CH_3 \\ \| \\ ClCH_2CH_2-N-CH_2CH_2Cl \end{array}$	Lung carcinoma and lung adenoma in mice Lung tumors in mice
BCNU (Alkylating agent)	$\begin{array}{c} NO \quad O \\ \| \quad \parallel \\ ClCH_2CH_2-N-C-NHCH_2CH_2Cl \end{array}$	Mammary carcinomas, kidney carcinomas, lung carcinomas, lymphosarcoma, brain tumors, etc Lung tumors, intra-abdominal tumor and skin tumor
CCNU (Alkylating agent)	$\begin{array}{c} O \quad NO \\ \parallel \quad \| \\ NH-C-N-CH_2CH_2Cl \end{array}$	Mammary carcinomas, kidney carcinomas, lung carcinomas, lymphosarcomas, leukaemias, skin tumors, brain tumors, etc Lung carcinomas
Azathioprine (Antimetabolite)		Bladder and kidney tumors in mice Thymic lymphomas, squamous cell carcinomas and ear-duct carcinoma in rats Carcinogen in man
Bromodeoxyuridine (Antimetabolite)		Neoplastic growth in the adult structures of Drosophila following treatment of larvae with bromodeoxyuridine in presence of 5-fluorouracil

310

Iododeoxyuridine
(Antimetabolite)

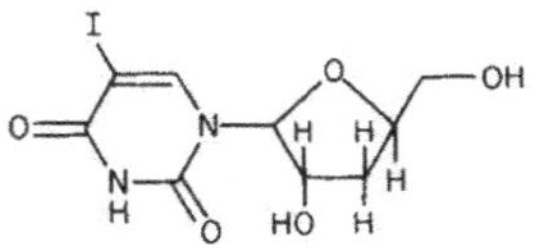

The same as bromodeoxyuridine

5-Fluorouracil
(Antimetabolite)

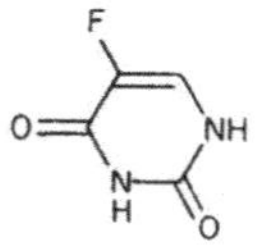

Co-carcinogen
5-Fluorouracil is mutagenic to RNA viruses such as tobacco mosaic and polio viruses

Carcinogen in Sprague-Dawley rats

Methotrexate
(Antimetabolite)

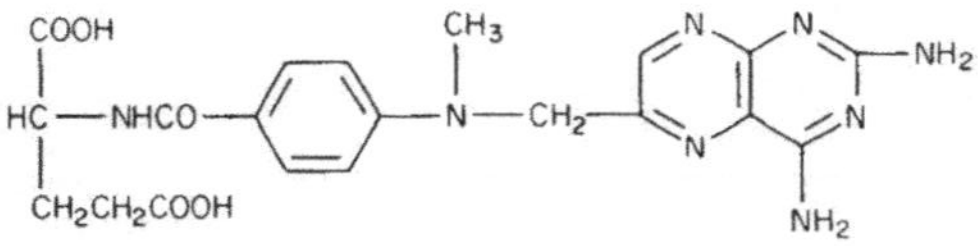

Carcinogen in Sprague-Dawley rats
Co-carcinogen   in Drosophila larvae

5-Azacytidine
(Antimetabolite)

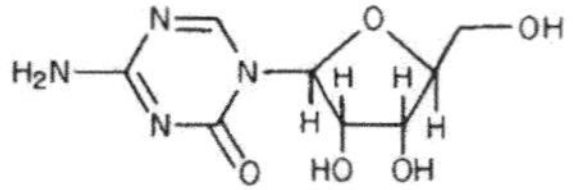

Inducing genetic mutations in *E. coli* and arbo-viruses producing well-characterized changes in the chromosomes of Vicia faba
Lymphocytic and granulocytic neoplasms of the hematopoietic system in rats

5-Mercaptopurine
(Antimetabolite)

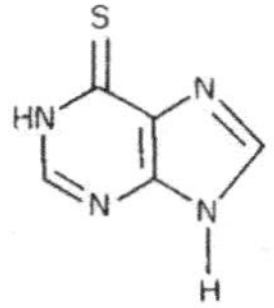

Causing increased incidences of certain tumors of hematopoietic system in rats and mice
Causing a high frequency of mispaired bases in DNA and consequent mutation in *E. coli*

TABLE 44 (Continued).

Name (Classification)	Structure	Carcinogenic activities

Actinomycin D (Antibiotic)

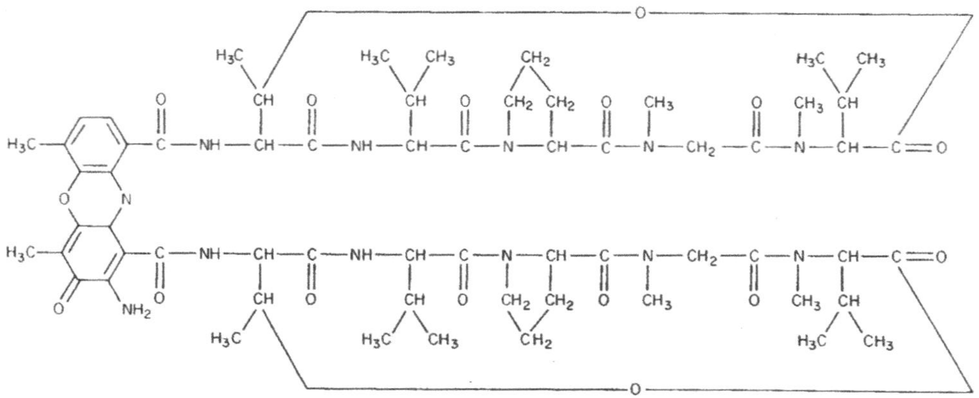

Mesotheliomas Peritoneal sarcomas in rats

Daunomycin (Antibiotic)

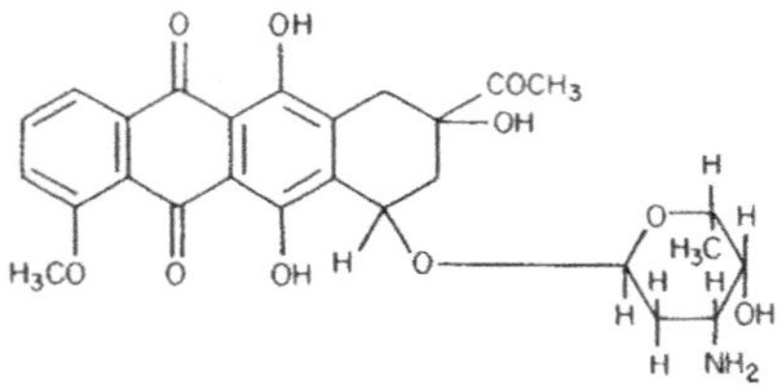

Nephrotoxic, inducing tumors of the kidneys and of the sex organ in rats
Causing fibrosarcomas in XVII/Rho mice

312

Mitomycin C
(Antibiotic)

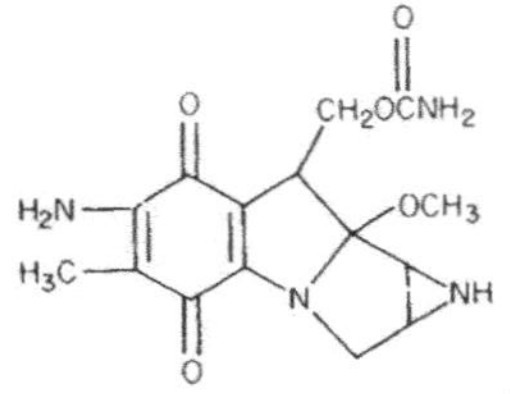

Peritoneal sarcomas in rats
Malignant, benign
Tumor in rats

Adriamycin
(Antibiotic)

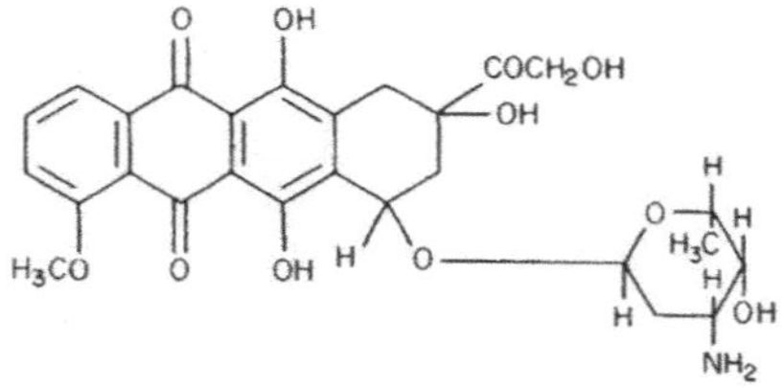

Inducing ocular and dental abnormalities and tumors in Charles River rats
A transforming agent

Colchicine
(Alkaloid)

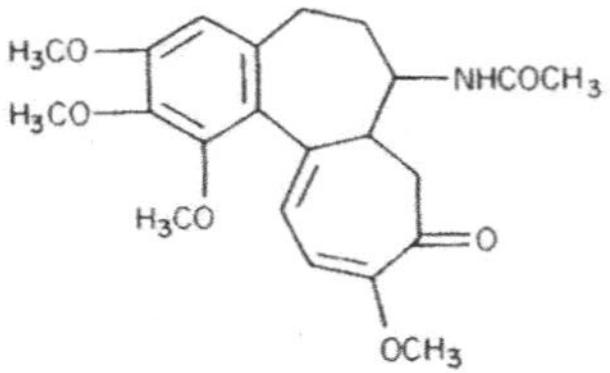

Skin tumors
Inhibiting the wheat coleoptile similar to that of cigarette smoke condensate

Diethylstilboestrol
(Hormone)

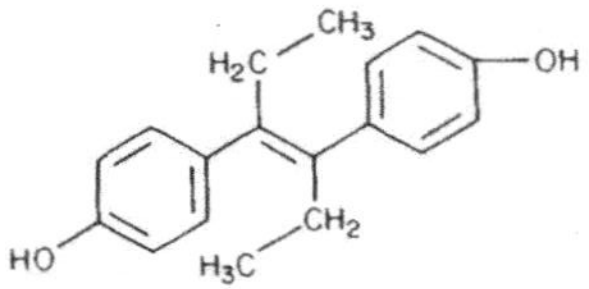

Vaginal cancer
Multiple cysts of the epididymis in male mice; tumors of the vagina and of the uterus in female mice
Fibrosarcomas   in female mice
Endometrial adenocarcinoma, ovarian adenocarcinoma in rats

TABLE 44 (Continued).

Name (Classification)	Structure	Carcinogenic activities
Ethinyloestradiol (Hormone)	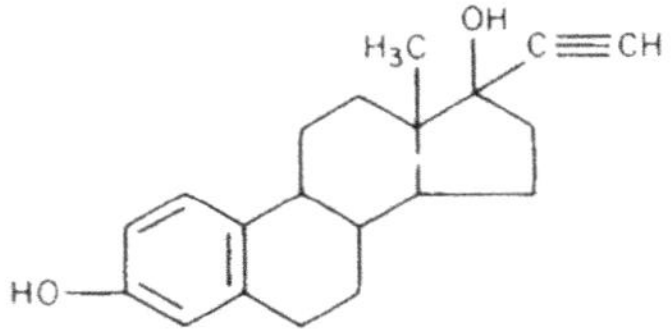	Liver tumours in female rats
Progestin (Hormone)	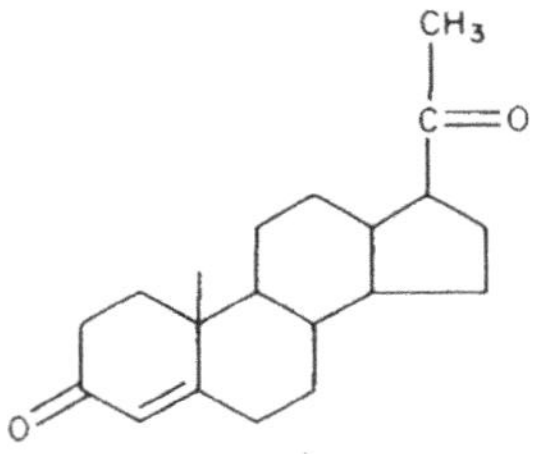	Ovarian tumors in Balb/c mice
Testosterone propionate (Hormone)	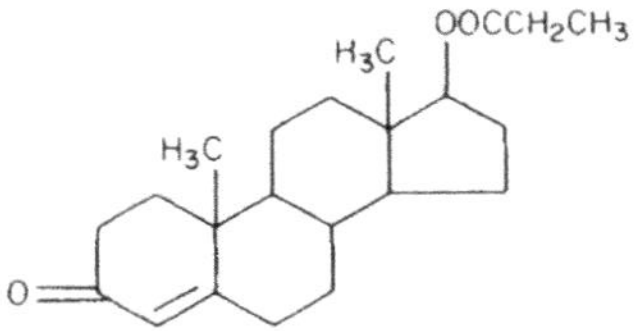	Uterine tumors in mice
Procarbazine (Miscellaneous)	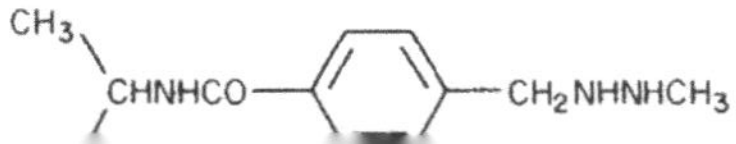	Acute myeloid leukemia and lymphomas A 30-fold higher risk of developing lymphoma than the control population

314

Compound	Structure	Findings
Hydroxyurea (Antimetabolite)	$\underset{H}{\overset{H}{N}}-\overset{\overset{O}{\|}}{C}-NH-OH$	Malignant transformation of hamster cells occurred after exposure to hydroxyurea
1-Acetyl-2-picolinoyl-hydrazine (Miscellaneous)	pyridine ring—$\overset{\overset{O}{\|}}{C}-\underset{H}{N}-\underset{H}{N}-COCH_3$	Lung tumors, ovarian and uterine tumors in mice, pancreatic tumors in rats
Phosphorodiamidic acid cyclohexylamine salt (Miscellaneous)	$OH\text{-}C_6H_{11}-NH_2 \quad \overset{\overset{O}{\|}}{P}(-NH_2)-NH_2$	Lung tumors, lymphomas, esophageal papillomas and uterine tumors in mice; bladder tumors, prostate tumors and subcutaneous tumors in rats
Cisplatin (Miscellaneous)	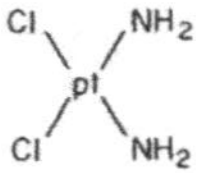 $Cl\text{—}Pt(NH_2)(NH_2)\text{—}Cl$	Epidermoid carcinomas, thymic lymphoma, pulmonary adenoma and skin papillomas Lung tumors, skin papillomas and carcinomas

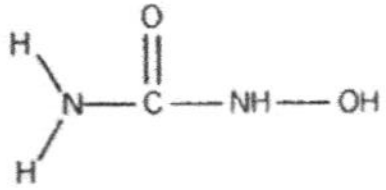

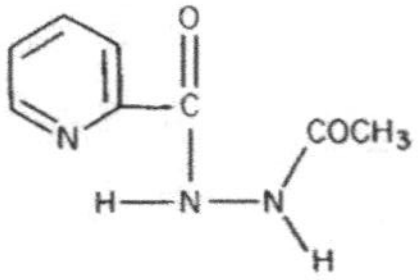

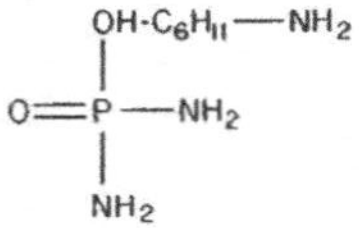

Source: Ref. 11, with permission from Blackwell Scientific Publications, London.

REFERENCES

1. *AMA Drug Evaluation*, 1st ed. (1971). American Medical Association, Chicago, Ill.

2. Grollman, A., and Grollman, E F. (1970). *Pharmacology and Therapeutics*, 7th ed. Lea & Febiger, Philadelphia, p. 396.

3. Lien, E. J., and Gudauskas, G. A. (1973). Structure side-effect of drugs. I. extrapyramidal syndrome. *J. Pharm. Sci. 62*: 645–647.

4. Lien, E. J. (1977). Adverse reactions and chemical structure, in H. Bundgaard, P. Juul and H. Kofod, Eds., *Drug Design and Adverse Reactions*, Alfred Benzon Symposium X: Drug Design and Adverse Reactions, Copenhagen, May, 1976, Munksgaard, Copenhagen, pp. 233–245.

5. Lien, E. J., and Lien, L. L. (1978). Structure side-effect sorting of drugs. III. Hepatotoxicities. *Calif. Pharm. 26*: 34–44.

6. Lien, E. J., and Lien, L. L. (1980). Structure side-effect sorting of drugs, IV: Nephrotoxicities. *J. Clin. Hosp. Pharm. 5*: 255–250.

7. Lien, E. J., Lipsett, L. R., and Lien, L. L. (1983). Structure side effect sorting of drugs. VI. Ototoxicities. *J. Clin. Hosp. Pharm. 8*: 15–33.

8. Lien, E. J., and Guduskas, G. A. (1973). Structure side-effect sorting of drugs. II. Skin sensitization. *J. Pharm. Sci. 62*: 1968–1971.

9. Lien, E. J., and Koda, R. T. (1981). Structure side-effect sorting of drugs. V. Glaucoma and cataracts associated with drugs and chemicals. *Drug. Intell. Clin. Pharm. 15*: 434–439.

10. Deguchi, T., Ishii, A., and Tanaka, M. (1978). Binding of aminoglycoside antibiotics to acidic mucopolysaccharides. *J. Antibiot. 31*: 150–155.

11. Lien, E. J., and Ou, X. C. (1985). Carcinogenicity of some anticancer drugs: a survey. *J. Clin. Hosp. Pharm. 10*: 223–242.

12. Dankner, R. (1984). Antihistamines, in P. E. Korenblat and H. J. Wedner, Eds., *Allergy, Theory and Practice.* Grune & Stratton, New York, pp. 205–219.

13. Cheng, H. C., and Woodward, J. K. (1982). Antihistaminic effect of terfenadine: a new piperidine type antihistamine. *Drug Dev. Res. 2:* 181–196.

14. Luscombe, D. K., Nicholls, P. J., and Parish, P. A. (1983). Comparison of the effects of azatadine maleate and terfenadine on human performance. *Pharmacotherapeutica 3:* 370–375.

15. Nicholson, A. N., Smith, P. A., and Spencer, M. B. (1982). Antihistamines and visual function: studies on dynamic acuity and the pupillary response to light. *Br. J. Clin. Pharmac. 14:* 683–690.

16. Krause, L. B., and Shuster, S. (1984). The effect of terfenadine on dermographic wealing. *Br. J. Dermatol. 110:* 73–79.

17. Camerman, A., and Camerman, N. (1970). Diphenylhydantoin and diazepam: molecular similarities and steric basis of anticonvulsant activity. *Science 168:* 1457–1458.

18. Jacobson, W. (1958). The toxic action of drugs on bone marrow, in A. L. Walpole and A. Spinks, Eds., *The Evaluation of Drug Toxicity.* Little, Brown, Boston, pp. 76–103.

19. Pisciotta, A. V. (1971). Drug-induced leukopenia and aplastic anemia. *Clin. Pharmacol. Ther. 12:* 13–43.

20. Davies, G. E. (1958). Allergic reactions as hazards in the use of new drugs, in A. L. Walpole and A. Spinks, Eds., *The Evaluation of Drug Toxicity.* Little, Brown, Boston, pp. 58–75.

21. Swanson, M., and Cook, R. (1977). *Drugs Chemicals and Blood Dyscrasias.* Drug Intelligence Publications, Inc., Hamilton, IL.

22. Cutting, W. C. (1969). *Handbook of Pharmacology,* 5th ed. Meredith, New York.

23. D'Arcy, P. F., and Griffin, J. P. (1972). *Iatrogenic Diseases.* Oxford University Press, Oxford.

24. Dipalma, J. R. Ed. (1971). *Drill's Pharmacology in Medicine*, 4th ed. McGraw-Hill, New York.

25. Goodman, L. S., and Gilman, A., Eds. (1970). *The Pharmacological Basis of Therapeutics*, 4th ed. Macmillan, Toronto, Canada.

26. Greenwald, E. S. (1970). *Cancer Chemotherapy*, 2nd ed. Medical Examination Publishing Co., Flushing, N. Y.

27. Grollman, A., and Grollma, E. F. (1970). *Pharmacology and Therapeutics*, 7th ed. Lea & Febiger, Philadelphia.

28. Holland, J. F., and Friei, E., III (1973). *Cancer Medicine*. Lea & Febiger, Philadelphia.

29. Krantz, J. C., Jr., and Carr, C. J. (1969). *The Pharmacological Principles of Medical Practice*, 7th ed. Williams & Wilkins, Baltimore.

30. Meyler, L., and Hexheimer, A. (1968). *Side Effects of Drugs*, Vol. VI. Excerpta Medica, Amsterdam.

31. Meyler, L., and Hexheimer, A. (1972). *Side Effects of Drugs*, Vol. VII. Excerpta Medica, Amsterdam.

32. Meyler, L., and Peck, H. M. (1968). *Drug Induced Diseases*, Vol. 3. Excerpta Medica, Amsterdam.

33. Moser, R. J. (1964). *Diseases of Medical Progress*, 2nd ed. Charles C Thomas, Springfield, Ill.

34. Moser, R. H. (1969). *Diseases of Medical Progress: A Study of Iatrogenic Disease*, Charles C Thomas, Springfield, Ill.

35. National Library of Medicine (1975). *Toxicity Bibliography*, Vol. 8. Bethesda, Md., No. 1—No. 4.

36. Peery, T. M., and Miller, F. N. (1971). *Pathology*, 2nd ed. Little, Brown, Boston.

37. Richards, D. J., and Rondel, R. K. (1972). *Adverse Drug Reactions*. Churchill Livingstone, Edinburgh.

38. Spain, D. M. (1963). *The Complications of Modern Medical Practices*. Grune & Stratton, New York.

39. Strecher, P. G., Ed. (1968). *The Merck Index*, 8th ed. Merck & Co., Rahway, N.J.

40. Hoagland, H. C. (1984). Hematologic complications of cancer chemotherapy, in M. C. Perry and J. W. Yarbro, Eds., *Toxicity of Chemotherapy*. Grune & Stratton, New York, pp. 433–448.

41. Davies, D. M. (1977). *Textbook of Adverse Drug Reactions*. Oxford University Press, Oxford.

42. Böttiger, L. E., Nordlander, M., Strandberg, I., and Westerholm, B. (1974). Deaths from drugs. An analysis of drug-induced deaths reported to the Swedish Adverse Drug Reaction Committee during a five-year period. *J. Clin. Pharmacol. 14*: 401–407.

43. Jack, D. B., Stenlake, J. B., and Templeto, R. (1972). Metabolism and excretion of guanoxan in man. *Xenobiotica 2*: 35–43.

44. Gall, E. A., and Mostofi, F. K., Eds. (1973). *The Liver*, Williams & Wilkins, Baltimore, Md.

45. Halsey, J. P., and Cardoe, N. (1982). Benoxaprofen: side-effect profile in 300 patients. *Br. Med. J. 284*: 1365–1369.

46. Carney, F. M. T., and Van Dyke, R. A. (1972). Halothane hepatitis: a critical review. *Anesth. Analg. Curr. Res. 51*: 135–160.

47. Holdsworth, C. D., Atkinson, M., and Goldie, W. (1961). Hepatitis caused by the newer amino-oxidase inhibiting drugs. *Lancet 2*: 621–623.

48. McDermott, W., Ormond, L., Muschenheim, C., Deuschle, K., McCune, R. M., Jr., and Tompsett, R. (1954). Pyrazinamide-isoniazid in tuberculosis. *Am. Rev. Tuberc. 69*: 319–333.

49. Kelsey, W. M., and Scharyj, M. (1967). Fatal hepatitis probably due to indomethacin. *J. Am. Med. Assoc. 199*: 586–587.

50. Bickers, J. N., Buechner, H. A., Hood, B. J., and Alvarez-Chiesa, G. (1961). Hypersensitivity reaction to antituberculosis drugs with hepatitis, lupus phenomenon and myocardial infarction. *N. Engl. J. Med. 265*: 131–132.

51. Maddrey, W. C., and Boitnott, J. K. (1975). Hepatitis

induced by isoniazid and methyldopa. *Hosp. Pract.*, April, 119–125.

52. Kuntz, E., Liehr, H., and Pfingst, R. (1967). Toxische Leberschadigun durch Athionamid. *Dtsch. Med. J. 92:* 1718–1722.

53. Schreiner, G. E., and Maher, J. F. (1965). Toxic nephropathy. *Am. J. Med. 38:* 409–449.

54. Dicker, S. E. (1970). *Mechanism of Urine Concentration and Dilution in Mammals.* Williams & Wilkins, Baltimore.

55. Appel, G. B., and Neu, H. C. (1977). The nephrotoxicity of antimicrobial agents (first of three parts). *N. Engl. J. Med. 296:* 663–669.

56. Appel, G. B., and Neu, H. C. (1977). The nephrotoxicity of antimicrobial agents (second of three parts). *N. Engl. J. Med. 296:* 722–728.

57. Appel, G. B., and Neu, H. C. (1977). The nephrotoxicity of antimicrobial agents (third of three parts). *N. Engl. J. Med. 296:* 784–787.

58. Zollinger, H. U., and Spühler, O. (1950). Die nicht-eirige, chronische interstitielle Nephritis. *Schweiz. Z. Allerg. Pathol. Bakteriol. 13:* 807–811.

59. Kincaid-Smith, P. (1978). Drug-induced renal disease. *Practitioner 220:* 862–867.

60. Kincaid-Smith, P. (1978). Analgesic nephropathy. *Kidney Int. 13:* 1–4.

61. Shelley, J. H. (1978). Pharmacological mechanisms of analgesic nephropathy. *Kidney Int. 13:* 15–26.

62. Murray, T. G., and Goldberg, M. (1978). Analgesic associated nephropathy in the U.S.A.: epidemiologic, clinical and pathogenic features. *Kidney Int. 13:* 64–71.

63. Dawson, A. G. (1975). Effects of Acetylsalicylate on Gluconeogenesis in isolated rat kidney tubules. *Biochem. Pharmacol. 24:* 1407–1411.

64. Tan, S. Y., Shapiro, P., and Kish, M. A. (1979). Reversible acute renal failure induced by indomethacin. *J. Am. Med. Assoc. 241:* 2732–2733.

65. Donker, A. J. M., Ariz, L., and Brentiens, J. R. (1975).
The effect of indomethacin on kidney function. *Kidney Int.* 7: 362.

66. Murray, G., and Von Stounasser, V. (1976). Glomerular
lesions in experimental renal papillary necrosis. *Br. J. Exp. Pathol.* 57: 23–29.

67. Gault, M. H., Blennerhasset, J., and Muehrucke, R. C.
(1971). Analgesic nephropathy: A clinicopathologic study
using electron microscopy. *Am. J. Med.* 151: 740–756.

68. Prescott, L. F. (1976). The nephrotoxicity of analgesics.
J. Pharm. Pharmacol. 18: 331–344.

69. Prescott, L. F. (1965). Effects of acetylsalicylic acid,
phenacetin, paracetamol and caffeine on renal tubular
epithelium. *Lancet ii:* 91–96.

70. Boy, E. M. (1959). The acute toxicity of caffeine. *Tocicol.
Appl. Pharmacol.* 1: 250–257.

71. Nanra, R. A., and Kincaid-Smith, P. (1970). Papillary
necrosis in rats caused by aspirin and aspirin containing
mixtures. *Br. Med. J.* 3: 559–561.

72. Mudge, G. H. (1965). Diuretics and other agents employed
in the mobilization of oedema fluid; xanthines, pyrimidines
and triazines in L. S. Goodman and A. Gilman, Eds., *The
Pharmacological Basis of Therapeutics.* MacMillan, New
York, p. 850.

73. Steele, T. W., and Edwards, K. D. G. (1971). Analgesic
nephropathy. Changes in various parameters of renal function following cessation of analgesic abuse. *Med. J. Aust.*
1: 181–187.

74. Nanra, R. W., and Kincaid-Smith, P. (1972). Chronic ef-
fect of analgesics on the kidney. In K. D. G. Edwards,
Ed., *Drugs Affecting Kideny Function and Metabolism,
Progress in Biochemical Pharmacology,* Vol. 7. Karger,
Basel, pp. 285–323.

75. Kalowski, S., Nanra, R. S., Mathew, T. H., and Kincaid-
Smith, P. (1973). Deterioration in renal function in association with co-trimoxazole therapy. *Lancet i:* 394–397.

76. Kutsher, A. H., Lane, S. L., and Segall, R. (1954). The
clinical toxicity of antibiotics and sulfonamides; a compara-

tive review of the literature based on 104, 672 cases treated systemically. *J. Allergy 25*: 135–150.

77. Lehr, D. (1945). Inhibition of drug precipitation in the urinary tract by the use of sulfonamide mixtures. I. Sulfathiazole-sulfadiazine mixture. *Proc. Soc. Exp. Biol. Med. 58*: 11–14.

78. Weinstein, L., Madoff, M. A., and Samet, C. A. (1960). The sulfonamides. *N. Engl. J. Med. 263*: 793–800.

79. Lien, E. J. (1979). The excretion of drugs in milk: A survey. *J. Clin. Pharm. 4*: 133–134.

80. Whelton, A., and Walker, W. G. (1974). Intrarenal antibiotic distribution in health and disease. *Kidney Int. 6*: 131–137.

81. Fillastre, J. P., Kuhn, J. M., Bendirdjian, J. P., Foucher, B., Leseur, J. P., Rollin, P., and Vaillant, R. (1976). Prediction of antibiotic nephrotoxicity. *Adv. Nephrol. 6*: 343–370.

82. Whelton, A., Sapir, D. G., Carter, G. G., Kramer, J., and Walker, W. G. (1971). Intrarenal distribution of penicillin, cephalothin, ampicillin and oxytetracycline during varied states of hydration. *J. Pharmacol. Exp. Ther. 179*: 419–428.

83. Bodey, G. P., Chang, H. Y., Rodriguez, V., and Stewart, D. (1975). Feasibility of administering aminoglycoside antibiotics by continuous intravenous infusion. *Antimicrob. Agents Chemother. 8*: 328–333.

84. Weinstein, L. (1975). Antimicrobial agents: penicillins and cephalosporins, in, L. S. Goodman and A. Gilman, Eds., *The Pharmacological Basis of Therapeutic*, 5th ed. Macmillan, New York, p. 1130.

85. Saslaw, S. (1970). Cephalosporins. *Med. Clin. North Am. 54*: 1217.

86. Balswin, D. S. Levin, B. B., McClusky, R. T., and Gallo, G. (1968). Renal failure and interstitial nephritis due to penicillin and methicillin. *N. Engl. J. Med. 279*: 1245–1252.

87. Burton, J. R., Lichtenstein, N. S., Calvin, R. B., and Hyslop, N. W., Jr. (1974). Acute renal failure during cephalothin therapy. *J. Am. Med. Assoc. 229*: 679–682.

88. Milman, N. (1978). Acute interstitial nephritis during treatment with penicillin and cephalothin. *Acta Med. Scand. 203*: 227–230.

89. Bear, D. M., Turk, M., and Petersdorf, R. G. (1970). Ampicillin. *Med. Clin. North Am. 54*: 1145–1159.

90. Gilbert, D., Gourley, M., D'Agostino R., and Goodnight, A. (1970). Interstitial nephritis due to methicillin, penicillin and ampicillin. *Ann. Allergy 28*: 378–385.

91. Mercer, C. W., Logic, J. R. (1973). Cardiac arrest due to hyperkalemia following intravenous penicillin administration. *Chest 64*: 358–359.

92. Lipner, H. J., Ruzany, F., Dasgupta, M., Lief, P. D., and Bank, N. (1975). The behavior of carbenicillin as a nonreabsorbable anion. *J. Lab. Clin. Med. 86*: 183–194.

93. Tattersall, M. H. N., Battersby, G., and Spiers, A. S. D. (1972). Antibiotics and hypokalemia. *Lancet 1*: 630–631.

94. Shils, M. E. (1963). Renal disease and the metabolic effects of tetracycline. *Ann. Inter. Med. 58*: 389–408.

95. Mavromatis, F. (1965). Tetracycline Nephropathy. *J. Am. Med. Assoc. 193*: 91–94.

96. Benitz, K. F., and Diermeier, H. G. (1964). Renal toxicity of tetracycline degradation products. *Proc. Soc. Exp. Biol. Med. 115*: 930–935.

97. Bennett, J. E., Brandriss, M. W., Butler, W. I., and Hill, G. J., II (1964). Amphotexicin B. toxicity. *Am. Intern. Med. 61*: 334–354.

98. Block, E. R., Bennett, J. E. Livoti, L. G., Klein, W. J., MacGregor, R. R., and Henderson, L. (1974). Flucytosine and amphotericin B hemodyalysis effects on the plasma concentration and clearance: studies in man. *Ann. Intern. Med. 80*: 613–617.

99. Feldman, H. A., Hamilton, J. D., and Gutman, R. A. (1973). Amphotericin B toxicity in anephric patient. *Antimicrob. Agents Chemother. 4*: 302–305.

100. Bennett, J. E. (1974). Chemotherapy of systemic mycosis. *N. Engl. J. Med. 290*: 30–32.

101. Butler, W. J. Bennett, J. E., Allen, D. W., Wertlake, P. T., Utz, J. P., and Hill, G. J., II (1964). Nephrotoxicity of amphotericin B early and late effects in 81 patients. *Ann. Intern. Med. 61:* 175–187.

102. Iovine, G., Berman, L. B., Halikis, D. N., Mowrey, F. H., Chappelle, G. H., and Gierson, H. W. (1963). Nephrotoxicity of amphotericin B: a clinical pathologic study. *Arch. Intern. Med. 112:* 853–862.

103. Hirsch, H. A., McCarthy, C. G., and Finland, M. (1960). Polymyxin B and colistin: activity, resistance and cross resistance in vitro. *Proc. Soc. Exp. Biol. 103:* 338–342.

104. Jawetz, E. (1951). Laboratory and clinical observations on polymyxin B and E. *Am. J. Med. 10:* 111–112.

105. Schoenbach, E. B., Bryer, M. S., and Lang, P. H. (1949). The clinical use of polymyxin. *Ann. NY Acad. Sci. 51:* 987–997.

106. Hoeprich, P. D. (1970). The polymyxins. *Med. Clin. North Am. 54:* 1257–1263.

107. Fekety, F., Norman, P., and Cluff, L. (1962). The treatment of gram-negative infections with colistin: The toxitity and efficacy of large doses in 48 patients. *Am. Inter. Med. 57:* 214–229.

108. Goodwin, N. J. (1970). Colistin and Sodium Colistimethate. *Med. Clin. North Am. 54:* 1267–1276.

109. Kunin, C. M., and Bugg, A. (1970). Recovery of tissue bound polymyxin B and colistimethate. *Proc. Soc. Exp. Biol. Med. 137:* 786–790.

110. Cluff, L. E., Caranosos, G. J., and Steart, R. B. (1975). *Clinical Problems with Drugs.* W. B. Saunders, Philadelphia.

111. Kirby, W. M. M. (1963). Vancomycin therapy of Staphylococcal infections. *Antibiot. Chemother. 11:* 84–96.

112. Riley, H. D. (1970). Vancomycin and Novobiocin. *Med. Clin. North Am. 54:* 1277–1289.

113. Setter, J. G., Haher, J. F., and Schreiner, G. E. (1963). Acute renal failure following cholescystography. *J. Am. Med. Assoc. 184:* 102–110.

114. Berk, R. N., Loeb, P. M., Goldberg, L. E., and Sokoloff, J. (1974). Oral cholecystograph with iopanoic acid. *N. Engl. J. Med. 290:* 204–210.

115. Krumlovsky, F. A., Simon, N., Santhanam, S., Dreco, F., del Roxe, D., and Pomaranc, M. M. (1978). Acute renal failure. *J. Am. Med. Assoc. 239*: 125—127.

116. Borra, S., Hawkins, D., Duguid, W., and Kaye, M. (1971). Acute renal failure and nephrotic syndrome after angiocardiography with melumine diatrizoate. *N. Engl.J. Med. 284*: 592—593.

117. Port, F. K., Wagoner, R. D., and Fulton, R. E. (1974). Acute renal failure after angiography. *Am. J. Roentgenol. Radium Ther. Nucl. Med. 121*: 544—550.

118. Ansari, Z., and Baldwin, D. S. (1976). Acute renal failure due to radio-contrast agents. *Nephron. 17*: 28—40.

119. MacEwan, D. W., and Dunbar, J. S. (1962). Intravenous pyelography in children with renal insufficiency. *Radiology 78*: 893—903.

120. Schwartz, W. B., Hurwit, A., and Ettinger, A. O. (1963). Intravenous urography in the patient with renal insufficiency. *N. Engl. J. Med. 269*: 277—283.

121. Fulton, R. E., Witten, D. M., and Waggoner, R. D. (1969). Intravenous urography in renal insufficiency. *Am. J. Roentgenol. Radium Ther. Nucl. Med. 106*: 623—634.

122. Light, J. A., and Hill, G. S. (1975). Acute tubular necrosis in a renal transplant recipient: complication from drip-infusion excretory urography. *J. Am. Med. Assoc. 232*: 1267—1268.

123. Postlethwaite, A. E., and Kelly, W. N. (1971). Uriscosuric effect of radio-contrast agents. *Ann. Intern. Med. 74*: 845—853.

124. Mudge, G. H. (1971). Uricosuric action of cholecystographic agents—A possible factor in nephrotoxicity. *N. Engl. J. Med. 284*: 929—933.

125. Friedman, I. (1974). *Pathology of the Ear.* Blackwell Scientific, Oxford, pp. 467—469.

126. Porter, J., and Jick, H. (1977). Drug-induced anaphylaxis, convulsions, deafness and extrapyramidal symptoms. *Lancet i*: 587—588.

127. Jackson, G. G., and Arcieri, G. (1971). Ototoxicity of gentamycin in man: a survey and controlled analysis of clinical experience in the United States. *J. Infect. Dis.* *124*, (Suppl.), 130–137.

128. Quick, C. A. (1976). Hearing loss in patients with dialysis and renal transplants. *Ann. Otol., Rhinol., & Laryngol. (St. Louis)* *85*: 776–790.

129. Quick, C. A., Chole, R. A., and Mauer, S. M. (1975). Deafness and renal failure due to potassium bromate poisoning. *Arch. Otolaryngol.* *101*: 494.

130. Robson, J., and Sullivan, F. (1963). Antituberculous drugs. *Pharmacol. Rev.* *15*: 169–223.

131. Straemmler, M., and Dudkowiak, V. (1961). Uber die wirkung des Kanamycin auf die nieren. *Med. Welt* *24*: 1296–1299.

132. Quick, C. A. (1976). Hearing loss in patients with dialysis and renal transplants. *Ann. Otol., Rhinol., & Laryngol. (St. Louis)* *85*: 776–790.

133. Buckley, J., and Hawthorne, J. (1972). Erythrocyte membrane polyphosphoinositide metabolism and the regulation of calcium binding. *J. Biol. Chem.* *247*: 7218–7223.

134. Kai, M., and Hawthorne, J. (1969). Physiological significance of polyphosphoinositides in brain. *Ann. N.Y. Acad. Sci.* *165*: 761–773.

135. Orsulakova, A., Stockhorst, E., and Schacht, J. (1976). Effect of neomycin on phosphoinositide labeling and calcium binding in guinea-pig inner ear tissues in vivo and in vitro. *J. Neurochem.* *26*: 285–290.

136. Schacht, J. (1976). Biochemistry of neomycin ototoxicity. *J. Accoust. Soc. Am.* *59*: 940–944.

137. Caspar, L. (1917). Zur Kenntnis der gewerblichen Augenschadigungen durch Naphthalin. *Klin. Monastbl. Augenheilkd.* *59*: 142.

138. Horner, W. D., Jones, R. B., and Boardman, W. W. (1935). Cataracts following the use of dinitrophenol. *J. Am. Med. Assoc.* *105*: 108.

139. Boardman, W. W. (1935). Rapidly developing cataract after dinitrophenol, *J. Am. Med. Assoc. 105*: 108.

140. Cogan, D. C., and Cogan, F. C. (1935). Dinitrophenol cataract. *J. Am. Med. Assoc. 105*: 793.

141. Lazar, N. K. (1935). Cataract following the use of dinitrophenol, *J. Am. Med. Assoc. 105*: 794.

142. Kniskern, P. W. (1935). Cataracts following dinitrophenol. *J. Am. Med. Assoc. 105*: 794.

143. Allen, T. D., and Benson, V. M. (1935). Late development of cataract following use of dinitrophenol about a year before. *J. Am. Med. Assoc. 105*: 795.

144. Shutes, M. H. (1935). Dinitrophenol. *Am. J. Opthalmol. 18*: 752.

145. Whalman, H. F. (1936). Dinitrophenol cataract. *Am. J. Ophthalmol. 19*: 885–888.

146. Hollwich, F., Boateng, A., and Kolck, B. (1975). In J. G. Bellow, Ed., *Cataract and Abnormalities of the Lens.* Grune & Stratton, New York.

147. Bron, A. J. (1979). Mechanisms of ocular toxicity, in J. W. Gorrod, *Drug. Toxicity.* Taylor & Francis, London, pp. 229–253.

148. Oppelt, W. W., White, E. D., and Halpert, E. S. (1969). The effect of corticosteroids on aqueous humor formation rate and outflow facility. *Invest. Ophthalmol. 8*: 535–541.

149. Reid, W. H., Blouin, P., and Schermer, M. (1976). A review of psychotropic medications and the glaucomas. *Int. Pharmacopsychiatry 11*: 163–176.

150. Berliner, M. L. (1939). Cataract following inhalation of paradichlorobenzene vapor. *Arch. Ophthalmol. 22*: 1023.

151. Von Sallmann, L., Grimes, P., and Collins, E. (1959). Mimosine cataract. *Am. J. Ophthalmol. 47*: 107–117.

152. Stempel, E., and Stempel, R. (1973). Drug-induced photosensitivity. *J. Am. Pharm. Assoc. NS 13*: 200–204.

153. FDA Drug Bulletin (1980). Food and Drug Administra-

tion, Department of Health and Human Services, Rockville, Md, *10* (No. 2).

154. Hanison, R. (1960). Bilateral lens opacities associated with the use of di-isopropyl fluorophosphate eyedrops. *Am. J. Ophthalmol. 50*: 153–154.

155. Axelsson, U., and Holmberg, Å. (1966). The frequency of cataract after miotic therapy. *Acta Ophthalmol. 44*: 421–428.

156. Shaffer, R. N., and Hetherington, J. (1966). Anticholinesterase drugs and cataracts. *Am. J. Ophthalmol. 62*: 613–618.

157. Fraunfelder, F. T. (1976). Drug-Induced Ocular Side Effects and Drug Interactions. Lea & Febiger, Philadelphia.

158. Becker, B. (1964). Cataracts and topical corticosteroids. *Am. J. Ophthalmol. 58*: 872–873.

159. Slansky, H. A., Kolbert, G., and Gartner, S. (1967). Exophthalmos induced by steroids. *Arch. Ophthalmol. 77*: 579–581.

160. Weinstock, M., Stewart, H. C., and Butterworth, K. R. (1958). Lenticular effect in mice of some morphine-like drugs. *Nature (London), 182*: 1519–1520.

161. Weinstock, M., and Stewart, H. C. (1961). Occurrence in rodents of reversible drug-induced opacities of the lens. *Br. J. Ophthalmol. 45*: 408–414.

162. Weinstock, M., and Scott, J. D. (1967). Effects of various agents on drug-induced opacities of the lens. *Exp. Eye Res. 6*: 368–375.

163. Weinstock, M., and Marshall, A. S. (1969). The influence of the sympathetic nervous system on the action of drugs on the lens. *J. Pharmacol. Exp. Ther. 166*: 8–13.

164. Chizek, D. J., and Franceschetti, A. T. (1969). Oral contraceptives: their side effects and ophthalmological manifestations. *Surv. Ophthalmol. 14*: 90–105.

165. Hollowich, F., and Verbeck, B. (1970). Side-effects of oral contraceptives on the eye. *Ger. Med. 15*: 155–159.

166. Varma, S. D., and Kinoshita, J. H. (1976). Inhibition of lens aldose reductase by flavonoids-their possible role

in the prevention of diabetic cataracts. *Biochem. Pharmacol. 25*: 2502–2513.

167. Segelman, A. B., Segelman, F. P., Varma, S. D., Wagner, H., and Seligmann, O. (1977). *Cannabis sativa* L. (marijuana) IX: lens aldose reductase inhibitory activity of marijuana flavone C-glysosides. *J. Pharm. Sci. 66*: 1358–1359.

168. Varma, S. D., Schocket, S. S., and Richards, R. P. (1978). Implications of aldose reductase in cataracts in human diabetes. *Invest. Ophthalmol. Visual Sci. 18*: 237–241.

169. Varma, S. D., Mikuni, I., and Kinoshita, J. H. (1975). Flavonoids as inhibitors of lens aldose reductase. *Science 188*: 1215–1216.

170. Rhoads, C. P. (1946). Nitrogen mustards in the treatment of neoplastic disease. *J. Am. Med. Assoc. 131*: 656–658.

171. Gilman, A. (1963). The initial clinical trial of nitrogen mustard. *Am. J. Surg. 105*: 574–578.

172. Li, M. C., Hertz, R., and Spencer, D. B. (1956). Effect of methotrexate therapy upon choriocarcinoma and chorioadenoma. *Proc. Soc. Exp. Biol. NY 93*: 361–366.

173. Hertz, R., Lewis, J., and Lipsett, M. B. (1961). Five years experience with chemotherapy of metastatic choriocarcinoma and related trophoblastic tumors in women. *Am. J. Obstet. Gynecol. 82*: 631–640.

174. De Vita, V. T., Jr., Henney, F. E., and Stonehill, E. (1979). Cancer mortality: the good news, in S. E. Jones and S. E. Salmon, Eds., *Adjuvant Therapy of Cancer II.* Grune & Stratton, New York, pp. xv–xx.

175. Perry, M. C., and Yarbro, J. W. (1984). Complciations of chemotherapy: an overview, in M. C. Perry and J. W. Yarbro, Eds., *Toxicity of Chemotherapy.* Grune & Stratton, New York, pp. 1–19.

176. Haskell, C. M., Ed. (1985). *Cancer Treatment*, 2nd ed., W. B. Saunders, Philadelphia.

177. Burger, A. (1983). *A Guide to the Chemical Basis of Drug Design.* Wiley, New York.

178. Mager, P. P. (1984). *Multidimensional Pharmacochemistry, Design of Safer Drugs.* Academic Press, New York.

179. Hansch, C., Li, R. L., Blaney, J. M., and Langridge, R. (1982). Comparison of the inhibition of *Escherichia coli* and *Lactobacillus casei* dihydrofolate reductase by 2,4-diamino-5-(substituted-benzyl)pyrimidines: quantitative structure-activity relationships, x-ray crystallography, and computer graphics in structure-activity analysis. *J. Med. Chem. 25:* 777–784.

180. Blaney, J. M., Hansch, C., Silipo, C., and Vittoria, A. (1984). Structure-activity relationships of dihydrofolate reductase inhibitors. *Chem. Rev. 84:* 333–407.

181. Seydel, J. K., Ed. (1985). *QSAR and Strategies in the Design of Bioactive Compounds.* Verlagsgesellschaft, Weinheim.

182. Hopfinger, A. J. (1985). Computer-assisted drug design. *J. Med. Chem. 28:* 1133–1139.

Appendix

Appendix I—Aromatic Group Dipole Moments [a]

No.	Formula	R	μ_R (Debye)	WLN[b]	Solvent[c]	Temperature, °C
1	Br	—Br	-1.57	*E	cHx	20
2	Cl	—Cl	-1.59	*G	B	25
3	F	—F	-1.43	*F	B	30
4	H	—H	0.03	*H	L	25
5	GeCl₃	—GeCl₃	-3.15	*-GE-GGG	B	25
6	I	—I	-1.36	*I	B	30
7	NO	—NO	-3.09[d]	*NO	B	25
8	NH₂	—NH₂	1.53	*Z	B	25
9	NO₂	—NO₂	-4.13	*NW	B	25
10	N₂H₃	—NHNH₂	1.80	*MZ	D	25
11	N₃	—N=N=N	-1.56	*NNN	B	25
12	OH	—OH	-1.59	*Q	B	25
13	PH₂	—PH₂	-1.11	*PHH	Hx	20
14	SFO₂	—SO₂F	-4.59	*SWF	B	25
15	SF₅	—SF₅	-3.44	*SFFFFF	D	25
16	SH	—SH	-1.33	*SH	B	25
17	SiCl₃	—SiCl₃	-2.40	*-SI-GGG	B	25
18	SiF₃	—SiF₃	-2.72	*-SI-FFF	B	25
19	CCl₃	—CCl₃	-2.03	*XGGG	CCl₄	25
20	CF₃	—CF₃	-2.61	*XFFF	B	25
21	CF₃O	—OCF₃	-2.36	*OXFFF	B	25
22	CF₃S	—SCF₃	-2.50	*SXFFF	B	25
23	CF₃Se	—SeCF₃	-2.48	*-SE-XFFF	B	ns
24	CN	—CN	-4.08	*CN	D	35
25	CNO	—N=C=O	-3.93	*NCO	B	20
26a	CNS	—SCN	-3.01	*SCN	B	25
26b	CNS	—NCS	-2.91	*NCS	B	20
27	CNSe	—SeCN	-4.01	*-SE-CN	B	25
28	CHO₂	—CHO	-3.02	*VH	B	25
29	CHO₂	—COOH	-1.30	*VQ	B	25
30	CHF₂O	—OCHF₂	-2.46	*OYFF	B	25
31	CHF₂S	—SCHF₂	-2.48	*SYFF	B	25
32	CHF₂OS	—SOCHF₂	-3.93	*SO&YFF	B	25
33	CHF₂O₂S	—SO₂CHF₂	-4.08	*SWYFF	B	25
34	CH₂Br	—CH₂Br	-1.87	*1E	B	25
35	CH₂Cl	—CH₂Cl	-1.83	*1G	B	20
36	CH₂I	—CH₂I	-1.60	*1I	CCl₄	25
37	CH₂NO	—CONH₂	-3.42	*VZ	B	25
38	CH₂NO	—CH=NOH(trans)	-0.87	*1UNQ-T	R	25

332

Appendix I—Aromatic Group Dipole Moments [a]

No.	Formula	R	μ_R (Debye)	WLN[b]	Solvent[c]	Temperature, °C
39	CH_2NO	—CH=NOH(cis)	-0.85	*1UNQ -C	B	25
40	CH_2NO	—NHCHO	-3.35	*MVH	CCl₄	ns
41	CH_2NO_2	—CH₂ONO	-2.10	*1ONO	ns	ns
42	CH_3	—CH₃	0.36	*1	B	25
43	CH_3O	—CH₂OH	1.73	*1Q	B	25
44	CH_3O	—OCH₃	-1.30	*O1	D	25
45	CH_3NS	—NHCSNH₂	-5.16	*MYZUS	B	25
46	CH_3OS	—SOCH₃	-3.98	*SO&1	B	20
47	CH_3O_2S	—SO₂CH₃	-4.75	*SW1	B	25
48	CH_3O_3S	—OSO₂CH₃	-3.77	*OSW1	B	25
49	CH_3S	—SCH₃	-1.34	*S1	ns	ns
50	CH_3Se	—SeCH₃	-1.31	*-SE-1	B	ns
51	CH_4N	—NHCH₃	1.69	*M1	D	25
52	CH_4NSO_2	—NHSO₂CH₃	-4.60	*MSW1	B	30
53	C_2H	—C≡CH	-0.77	*1UU1	D	25
54	C_2H_2N	—CH₂CN	-3.60	*1CN	D	25
55	$C_2H_2N_2O_2$	(structure)	-6.63	*AT5NNVOJ	B	30
56	C_2H_3	—CH=CH₂	0.20	*1U1	B	25
57	C_2H_3O	—COCH₃	-2.90	*V1	B	25
58	$C_2H_3O_2$	—OCOCH₃	-1.72	*OV1	B	25
59	$C_2H_3O_2$	—COOCH₃	-1.92	*VO1	D	25
60	$C_2H_3O_2$	—CH₂COOH	-1.86	*1VQ	D	25
61	C_2H_4NO	—NHCOCH₃	-3.65	*MV1	B	25
62	C_2H_4NS	—NHCSCH₃	-4.28	*MYUS	cHx	25
63	C_2H_5	—CH₂CH₃	0.39	*2	B	25
64	C_2H_5O	—OC₂H₅	-1.38	*O2	B	25
65	$C_2H_5O_2S$	—SO₂C₂H₅	-3.48	*SW2	D	25
66	$C_2H_5O_3S$	—SO₃C₂H₅	-4.99	*SW02	D	25
67	C_2H_5S	—CH₂SCH₃	1.46	*1S1	D	25
68	C_2H_5S	—SC₂H₅	-4.08	*S2	B	25
69	C_2H_6N	—N(CH₃)₂	1.61	*N1&1	B	25
70	C_2H_6OP	—PO(CH₃)₂	-4.39	*PO&1&1	B	20
71	C_2H_6P	—P(CH₃)₂	-1.31	*P1&1	B	20
72	C_2F_3	—C≡CCF₃	-3.38	*1UU1XFFF	B	ns
73	C_3F_7	—CF(CF₃)₂	-2.68	*XFXFFFXFFF	B	25
74	C_3HF_6O	—C(OH)(CF₃)₂	-1.71	*XQXFFFXFFF	L	25

continued

Appendix I—*Continued*

No.	Formula	R	μ_R (Debye)	WLN[b]	Solvent[c]	Temperature, °C
75	$C_3H_2F_3$	—CH=CHCF$_3$	-2.79	*1U1XFFF	B	ns
76	$C_3H_2F_3$	—C(CF$_3$)=CH$_2$	-2.25	*YU1&XFFF	B	20
77	C_3H_2N	—CH=CHCN(*trans*)	-4.12	*1U1CN –T	B	20
78	C_3H_2N	—CH=CHCN(*cis*)	-3.54	*1U1CN –C	B	20
79	C_3H_2NS	(2-methylthiazol-5-yl)	-1.21	*BT5N CSJ	CCl$_4$	20
80	C_3H_2NS	(4-methylthiazol-2-yl)	-1.33	*ET5N CSJ	CCl$_4$	20
81	C_3H_2NS	(5-methylthiazol-2-yl)	-1.89	*DT5N CSJ	CCl$_4$	20
82	$C_3H_2O_2$	—COCH$_2$CO—	-2.73	*V1V*	D	25
83	$C_3H_2O_2$	—OCOCH=CH—	-4.63	*OV1U1*	D	25
84	C_3H_3O	—CH=CHCHO	-2.71	*1U1VH	B	25
85	$C_3H_3O_2$	—CH=CHCOOH	-2.04	*1U1VQ	B	ns
86	$C_3H_3O_2$	—COCOCH$_3$	-2.44	*VV1	B	25
87	$C_3H_3N_2$	(1-imidazolyl)	-3.14	*AT5N CNJ	B	25
88	$C_3H_3N_2$	(pyrazolyl)	-2.00	*AT5NNJ	B	25
89	$C_3H_3N_2$	(methyl-pyrazolyl)	-2.26	*CT5MNJ	B	25
90	$C_3H_3N_2O$	(hydroxy-methyl-pyrazolyl)	-2.18	*AT5NNJ CQ	D	25

Appendix I—*Continued*

No.	Formula	R	μ_R (Debye)	WLN[b]	Solvent[c]	Temperature, °C
91	$C_3H_3N_2O$		-2.43	*AT5NNJ DQ	D	25
92	$C_3H_3N_2O$		-3.41	*AT5NNJ EQ	D	25
93	C_3H_3Se	—SeC≡CCH₃	-1.31	*-SE-1UU2	B	25
94	$C_3H_3O_2Se$	—SeCH=CHCOOH(*trans*)	-2.27	*SE-1U1VQ –T	B	25
95	$C_3H_3O_2Se$	—SeCH=CHCOOH(*cis*)	-1.69	*-SE-1U1VQ –C	B	25
96	C_3H_4	—CH=CHCH₂—	0.62	*1U2*	B	25
97	$C_3H_4O_2$	—CH₂CH₂COO—	-3.85	*2VO*	B	25
98	C_3H_5		0.51	*AL3TJ	B	25
99	C_3H_5O	—COC₂H₅	-2.90	*V2	B	30
100	$C_3H_5O_2$	—CH₂OCOCH₃	-1.68	*1OV1	L	25
101	$C_3H_5O_2$	—COOC₂H₅	-1.85	*VO2	B	25
102	$C_3H_5O_2$	—CH₂COOCH₃	-1.81	*1VO1	B	24
103	$C_3H_5O_2$		-1.97[e]	*BT5O COTJ	B	20
104	C_3H_5OS		-1.30	*O– CT4STJ	ns	ns
105	C_3H_5OS	—COSC₂H₅	-1.55	*VS2	B	25
106	C_3H_5OS	—CSOC₂H₅	-2.24	*YUS&O2	B	25
107	C_3H_6	—CH₂CH₂CH₂—	0.55	*3*	cHx	25
108	C_3H_6	—C(CH₃)=CH₂	0.77	*YU1	B	25
109	C_3H_6NO	—N(CH₃)COCH₃	-3.60	*N1&V1	B	25
110	C_3H_7	—CH(CH₃)₂	0.40	*Y	cHx	25
111	$C_3H_7O_2$	—CH(OCH₃)₂	-1.06[e]	*YO1&O1	B	20
112	C_4F_9	—(CF₂)₃CF₃	-2.86	*/XFF/4F	B	25
113	C_4H_3S		0.81	*BT5SJ	B	25

335

Appendix I—Continued

No.	Formula	R	μ_R (Debye)	WLN[b]	Solvent[c]	Temperature, °C
114a	C_4H_4	—(CH)$_4$—	0.00	*RA*B*	B	25
114b	C_5H_5O	—CH=CHCOCH$_3$ (cis)	−2.89	*1U1V1	B	25
115	C_4H_8	—(CH$_2$)$_4$—	0.73	*4*	B	25
116	$C_4H_7O_2$	—CH$_2$OCOC$_2$H$_5$	−1.80	*1OV2	L	28
117	$C_4H_7O_2$	—CH$_2$COOC$_2$H$_5$	−1.85	*1VO2	B	24
118	$C_4H_7O_2$	—COOCH(CH$_3$)$_2$	−1.82	*VOY	B	25
119	$C_4H_7O_2$	—CH$_2$CH$_2$OCOCH$_3$	−1.86	*2OV1	B	25
120	$C_4H_5O_2$	—CH=CHCOOCH$_3$	−2.13	*1U1VO1	B	30
121	$C_4H_7O_2$		−1.47[e]	*BT6O COTJ	B	20
122	C_4H_9	—C(CH$_3$)$_3$	0.52	*X	B	25
123	C_4H_9O	—O(CH$_2$)$_3$CH$_3$	−1.19	*O4	B	20
124	$C_4H_{10}O_3P$	—PO(OC$_2$H$_5$)$_2$	−3.04	*PO&O2&O2	CCl$_4$	25
125	$C_4H_{11}Si$	—CH$_2$Si(CH$_3$)$_3$	0.68	*1-SI-1&1&1	B	25
126	C_5N_3	—C(CN)=C(CN)$_2$	−5.30	*YCN&UYCN&CN	B	30
127	C_5H_3O		−4.10	*CL5VJ	B	30
128	C_5H_3OS		−3.45	*V–BT5SJ	B	25
129	$C_5H_3S_2$		−3.15	*YUS–BT5SJ	B	ns
130	C_5H_4N		−1.94	*BT6NJ	B	25
131	C_5H_4N		−2.28	*CT6NJ	B	25
132	C_5H_4N		−2.57	*DT6NJ	B	25
133	C_5H_4NO		−1.64	*V–BT5MJ	B	25

336

Appendix I—*Continued*

No.	Formula	R	μ_R (Debye)	WLN[b]	Solvent[c]	Temperature, °C
134	C_5H_4NO		−4.52	* DT6NJ AO	ns	ns
135	C_5H_4NO	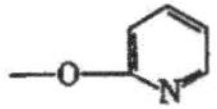	−1.96	*O–BT6NJ	B	20
136	C_5H_4NO	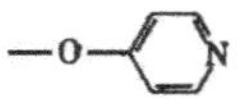	−2.46	*O–DT6NJ	B	20
137	$C_5H_4NO_3$	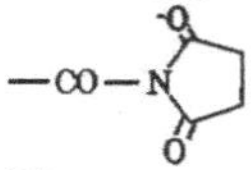	−4.10	*V–AT5NTJ	D	30
138	C_5H_5S	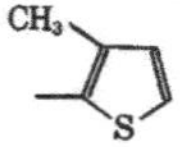	−1.10	* BT5SJ Cl	B	25
139	C_5H_5S	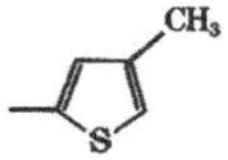	−0.88	* BT5SJ D1	B	25
140	C_5H_5S	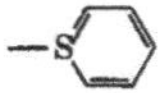	−0.79	* AT6SJ &5	B	30
141	$C_5H_6NO_2$	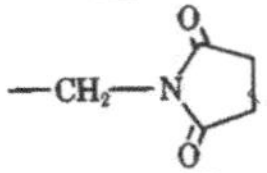	1.80	*1–AT5NTJ	D	25
142	$C_5H_6NO_2$	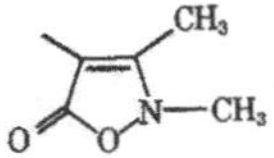	−5.65	* DT5NOVTJ A1 E1	B	ns
143	$C_5H_6NO_2$	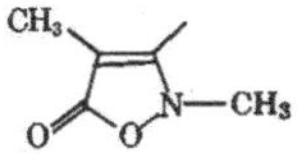	−5.70	* ET5NOVTJ A1 D1	B	ns

continued

Appendix I—*Continued*

No.	Formula	R	μ_R (Debye)	WLN[b]	Solvent[c]	Temperature, °C
144	$C_5H_6NO_2$	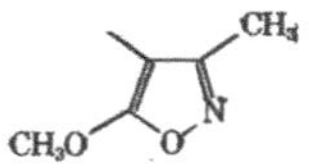	−2.32	* DT5NOJ CO1 E1	D	ns
145	$C_5H_6NO_2$	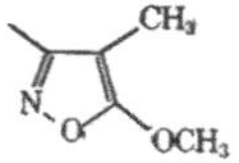	−2.83	* ET5NOJ CO1 D1	D	ns
146	$C_5H_7N_2O$	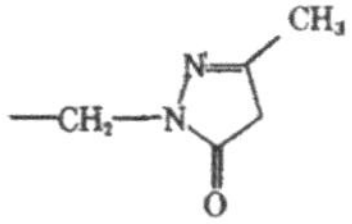	−3.10	*1- BT5NNV DHJ E1	D	25
147	$C_5H_7N_2O$	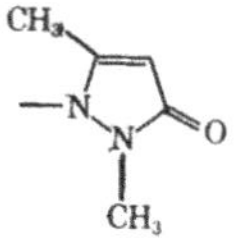	−5.47	* AT5NNVJ B1 E1	D	25
148	$C_5H_7N_2O$	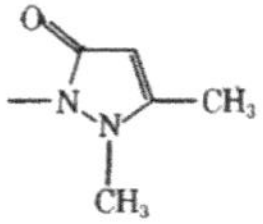	−5.47	* BT5NNVJ A1 E1	B	25
149	$C_5H_7N_2S$	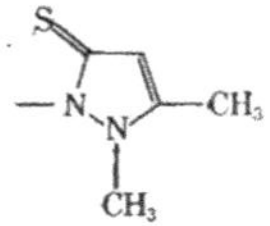	−7.60	* BT5NNYJ A1 CUS E1	D	25
150	$C_5H_7N_2S$	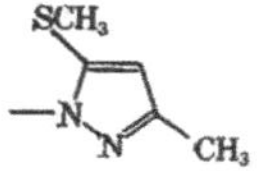	−2.80	* BT5NNJ CS1 E1	B	25
151	$C_5H_7O_2$	$-CH=CHCOOC_2H_5$	−1.73	*1U1VO2	L	26

Appendix I—*Continued*

No.	Formula	R	μ_R (Debye)	WLN[b]	Solvent[c]	Temperature, °C
152	C_5H_8NO	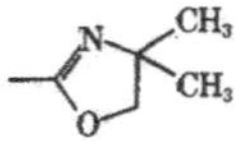	−1.11	* BT5N CO AUTJ E1 E1	B	25
153	C_5H_8NO	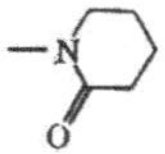	−3.96	* AT6NVTJ	B	25
154	$C_5H_8NO_3$	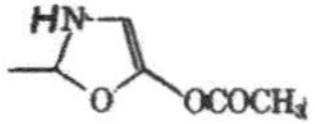	−2.95	* BT5N COJ DOV1	B	ns
155	$C_5H_8NO_4$	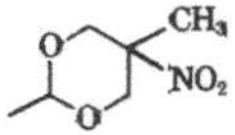	−4.47	* BT6O COTJ ENW E1	B	ns
156	$C_5H_{11}O_2$	$-CH(OC_2H_5)_2$	−1.23[e]	*YO2 &O2	B	20
157	$C_6H_4N_2Br$	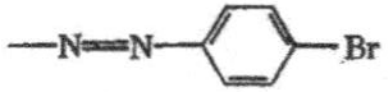	−1.47	*NUNR DE	B	25
158	C_6H_4BrO	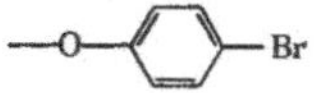	−1.59	*OR DE	B	20
159	C_6H_4BrO	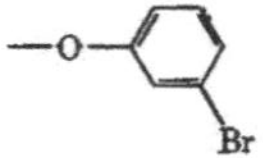	−1.78	*DR CE	B	20
160	C_6H_4BrO	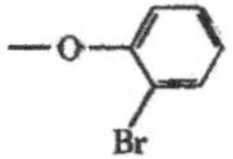	−2.20	*OR BE	B	20
161	$C_6H_4BrO_3S$	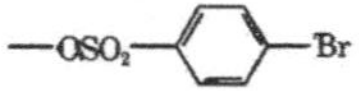	−3.82	*OSWR DE	B	25
162	$C_6H_4N_2Cl$	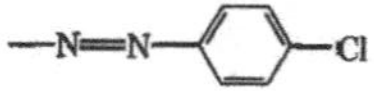	−1.56	*NUNR DG	B	25
163	C_6H_4I		−1.55	*R DI	B	25

Appendix I—*Continued*

No.	Formula	R	μ_R (Debye)	WLN[b]	Solvent[c]	Temperature, °C
164	C_6H_4IO		−2.06	*OR BI	B	20
165	C_6H_4IO		−1.68	*OR CI	B	20
166	C_6H_4IO		−1.47	*OR DI	B	20
167	C_6H_4IS		−2.38	*SR BI	B	20
168	C_6H_4IS		−1.80	*SR CI	B	20
169	C_6H_4IS		−1.50	*SR DI	B	20
170	C_6H_4NO		−2.95	*V– BT6NJ	B	25
171	C_6H_4NO		−3.01	*V– CT6NJ	B	25
172	C_6H_4NO		−3.06	*V– DT6NJ	B	25
173	$C_6H_4NO_2S$		−5.22	*SR BNW	B	20
174	$C_6H_4NO_2S$		−4.04	*SR CNW	B	20

340

Appendix I—*Continued*

No.	Formula	R	μ_R (Debye)	WLN[b]	Solvent[c]	Temperature, °C
175	$C_6H_4NO_3$	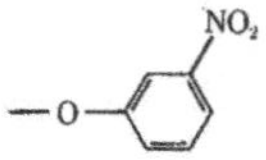	−4.04	*OR CNW	B	20
176	$C_6H_4NO_3$	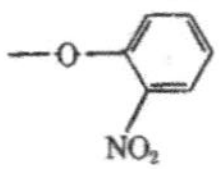	−4.60	*OR BNW	B	20
177	$C_6H_4NO_5S$	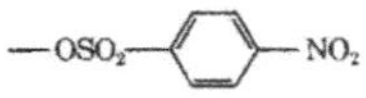	−4.72	*OSWR DNW	B	25
178	$C_6H_4NO_5S$	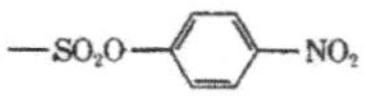	−2.76	*SWOR DNW	B	ns
179	$C_6H_4N_3O_4$	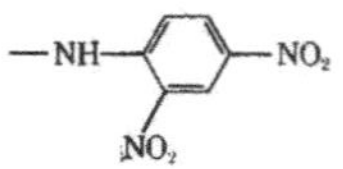	−6.36	*MR BNW DNW	B	ns
180	C_6H_5	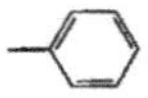	0.00	*R	L	ns
181	C_6H_6NO	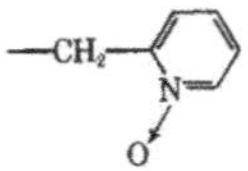	−4.13	*1– BT6NJ AO	B	25
182	C_6H_6NO	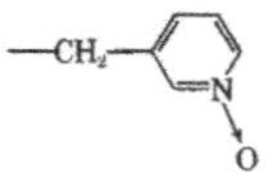	−4.61	*1– CT6NJ AO	B	25
183	C_6H_6NO	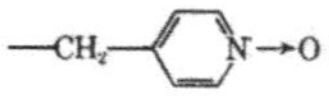	−4.63	*1– DT6NJ AO	D	25
184	$C_6H_5N_2$	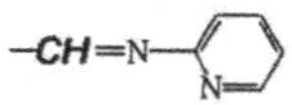	−1.36	*1UN– BT6NJ	B	25
185	$C_6H_5N_2$	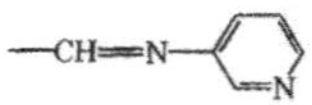	−2.98	*1UN– CT6NJ	B	25

continued

341

Appendix I—*Continued*

No.	Formula	R	μ_R (Debye)	WLN[b]	Solvent[c]	Temperature, °C
186	$C_6H_5N_2$	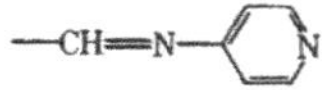 —CH=N— (pyridyl)	−4.16	*1UN– DT6NJ	B	25
187	$C_6H_5N_2O$	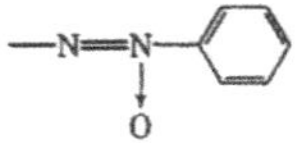 —N=N—→O (phenyl)	−1.73	*NUNO&R	B	25
188	$C_6H_5N_2O$	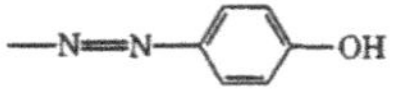 —N=N— (phenol)	−1.66	*NUNR DQ	B	25
189	C_6H_5O	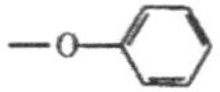 —O— (phenyl)	−1.16	*OR	B	25
190	C_6H_5O	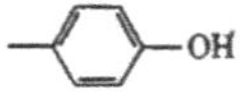 (hydroxyphenyl, OH)	−1.34	*R DQ	B	20
191	C_6H_5O	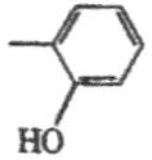 HO (phenyl)	−1.63	*R BQ	B	25
192	C_6H_5OS	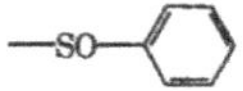 —SO— (phenyl)	−4.07	*SO&R	B	25
193	$C_6H_5O_2S$	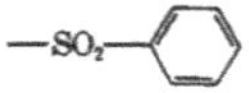 —SO₂— (phenyl)	−5.05	*SWR	B	25
194	$C_6H_5O_3S$	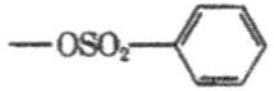 —OSO₂— (phenyl)	−4.72	*OSWR	B	25
195	C_6H_5S	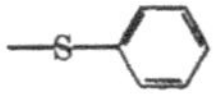 —S— (phenyl)	1.55[e]	*SR	B	25
196	$C_6H_5S_2$	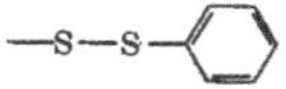 —S—S— (phenyl)	1.79	*SSR	B	25
197	C_6H_6N	—NHC₆H₅	1.11	*MR	B	25
198	C_6H_6N	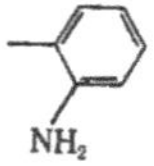 NH_2 (aminophenyl)	1.45	*R BZ	B	25

Appendix I—Continued

No.	Formula	R	μ_R (Debye)	WLN[b]	Solvent[c]	Temperature, °C
199	C_6H_6N	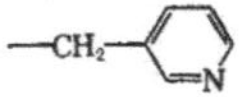	−2.18	*1-CT6NJ	B	25
200	C_6H_6N	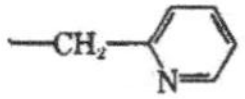	−1.89	*1-BT6NJ	B	25
201	C_6H_6N	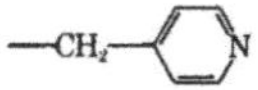	−2.65	*1-DT6NJ	B	25
202	$C_6H_6NO_2S$	$-NHSO_2C_6H_5$	−4.58	*MSWR	B	25
203	C_6H_6NS	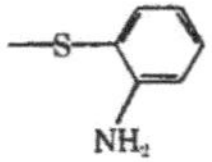	−1.87	*SR BZ	B	20
204	C_6H_6NS		−2.44	*SR DZ	B	20
205	$C_6H_6N_3$	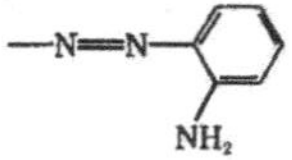	1.49	*NUNR BZ	B	20
206	$C_6H_6N_3$	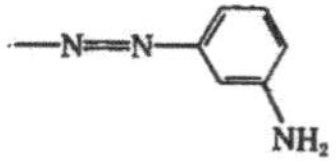	1.71	*NUNR CZ	B	20
207	$C_6H_6N_3$	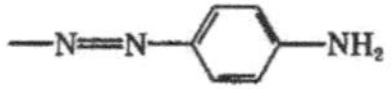	2.50	*NUNR DZ	B	20
208	$C_6H_9N_2O$	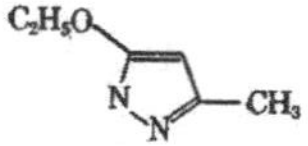	−2.65	* AT5NNJ C1 EO2	D	25
209	$C_6H_9N_2O$	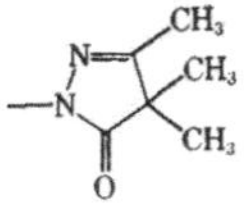	−2.83	* BT5NNV DHJ D1 D1 E1	D	25

343

Appendix I—*Continued*

No.	Formula	R	μ_R (Debye)	WLN[b]	Solvent[c]	Temperature, °C
210	$C_6H_9N_2S$	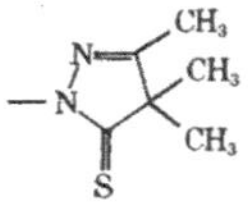	−3.16	*BT5NNY DHJ CUS D1 D1 E1	D	25
211	C_6H_{11}	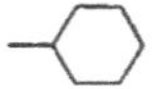	0.62	*AL6TJ	B	30
212	$C_6H_{11}O$	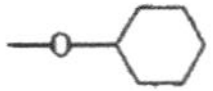	−1.55	*O–AL6TJ	B	20
213	$C_6H_{11}O$	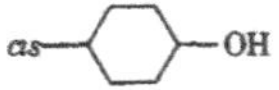 *cis*	−1.87	*AL6TJ DQ –C	B	25
214	$C_6H_{11}O$	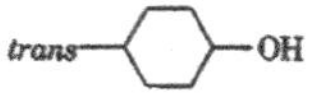 *trans*	−1.56	*AL6TJ DQ –T	B	25
215	$C_6H_{11}O_2$	$-CH_2CH_2OCOC_3H_7$	−1.85	*2OV3	L	25
216	$C_6H_{11}O_2$	$-COOC_5H_{11}-n$	−1.99	*VO5	L	25
217	$C_6H_{14}N$	$-N(i-C_3H_7)_2$	1.53	*NY&&Y	B	25
218a	C_7H_4NO	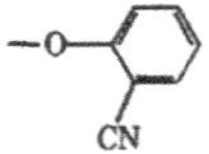	−4.88	*OR BCN	B	20
218b	C_7H_4NO	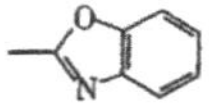	−1.22	*CT56 BN DOJ	B	25
219	C_7H_4NO	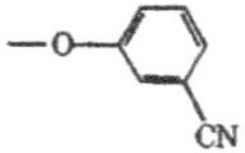	−4.01	*OR CCN	B	20
220	C_7H_4NO	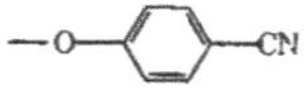	−4.23	*OR DCN	B	20
221	$C_7H_4NO_4$	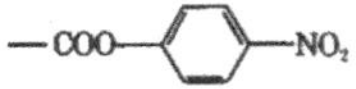	−4.43	*VOR DNW	B	40
222	C_7H_4NS	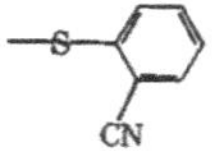	−5.04	*SR BCN	B	20

344

Appendix I—*Continued*

No.	Formula	R	μ_R (Debye)	WLN[b]	Solvent[c]	Temperature, °C
223a	C_7H_4NS	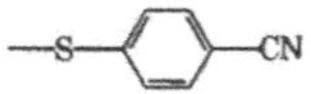 (4-cyanophenyl–S–)	−4.14	*SR DCN	B	20
223b	C_7H_4NS	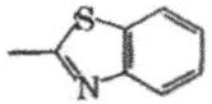 (2-benzothiazolyl)	−0.94	* CT56 BN DSJ	B	20
224	$C_7H_5O_2$	$-OCOC_6H_5$	−1.90	*OVR	B	25
225	C_7H_5O	$-COC_6H_5$	−3.04	*VR	B	25
226	$C_7H_5O_3$	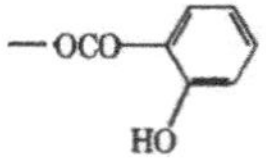 (2-hydroxyphenyl–OCO–, HO)	−1.92	*OVR BQ	B	25
227a	C_7H_6N	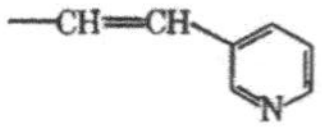 ($-CH=CH$ pyridyl)	−2.90	*1U1–CT6NJ	B	25
227b	C_7H_6N	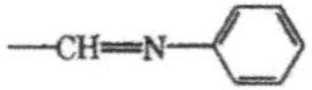 ($-CH=N$ pyridyl)	−1.61	*YUNR	B	25
228a	C_7H_6N	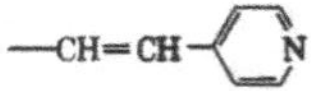 ($-CH=CH$ pyridyl)	−2.70	*1U1–DT6NJ	B	25
228b	C_7H_6N	 ($-N=CH$ phenyl)	1.61	*NUYR	B	25
229	C_7H_6NO	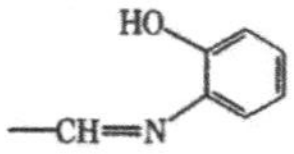 (HO, $-CH=N$)	−2.73	*1UNR BQ	B	25
230	C_7H_6NO	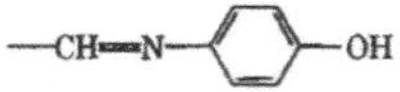 ($-CH=N$, OH)	−1.94	*NUCUNR DQ	B	25
231a	C_7H_6NO	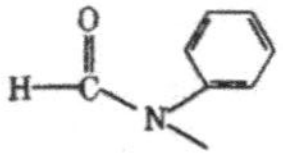 (N-methyl-N-phenylformamide)	−3.44	*NR&VH	B	25
231b	C_7H_7	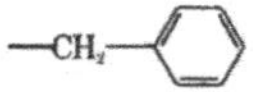 ($-CH_2$ phenyl)	0.36	*1R	B	20

continued

No.	Formula	R	μ_R (Debye)	WLN[b]	Solvent[c]	Temperature, °C
232	C_7H_7BrNO		−4.21	* DL6NTJ BE DCN	B	25
233	$C_7H_7N_2$		−2.03	*1UNMR	B	25
234	$C_7H_7N_2$		−2.20	*NR&YUM	D	25
235	$C_7H_7N_2O$		−1.54	*NUNR DO1	B	20
236	C_7H_7O		−1.38	*R BO1	B	35
237	$C_7H_7O_2$		−1.16	*O1OR	B	25
238	$C_7H_7O_3S$		−5.29	*OSWR D1	B	ns
239	$C_7H_7S_2$		−1.34	*S1SR	B	25
240	C_7H_8N		1.24	*N1&R	B	20
241	C_7H_8N		1.84	*1R DZ	B	ns
242	C_7H_8NO		−3.63	* DL6VTJ DCN	B	30
243	C_7H_8NO		−1.79	*MR C01	B	25

Appendix I—*Continued*

No.	Formula	R	μ_R (Debye)	WLN[b]	Solvent[c]	Temperature, °C
244	$C_7H_8NO_2S$	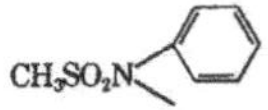	−4.41	*NR&SW1	D	30
245	$C_7H_8NO_3S$	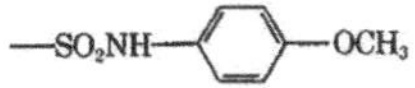	−5.08 −5.44	*SWMR DO1	B D	ns 25
246	$C_7H_8NO_3S$	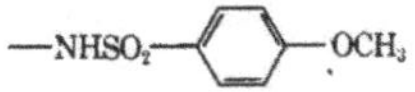	−5.21 −5.65	*MSWR DM1	B D	ns 25
247	$C_7H_8N_3$	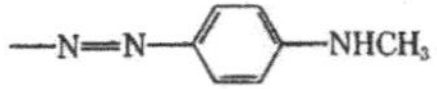	2.91	*NUNR DM1	B	20
248	C_7H_8P	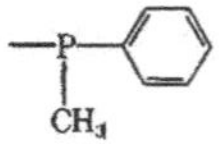	−1.39	*P1&R	B	20
249	$C_7H_9N_2O$	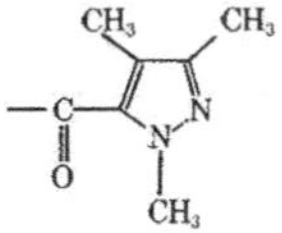	−2.80	*V−CT5NNJ B1 D1 E1	B	25
250	C_7H_9O	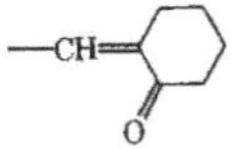	−3.23	*1U BL6VYTJ	B	25
251	$C_7H_{12}NO_2$	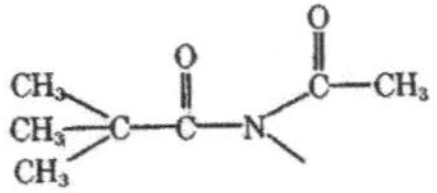	−3.59	*NV1&VX	B	20
252	$C_7H_{13}O_2$	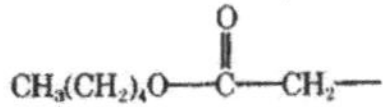	−2.13	*1VO5	B	25
253	$C_7H_{13}O_2$	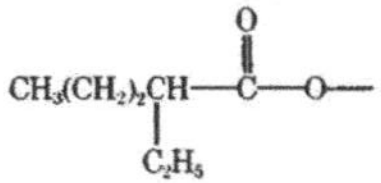	−2.05	*OVY2&3	B	25

347

Appendix I—*Continued*

No.	Formula	R	μ_R (Debye)	WLN[b]	Solvent[c]	Temperature, °C
254a	$C_7H_{13}O_2$		−1.70	*BT5O COTJ D1 D1 E1 E1	B	25
254b	C_8H_5	$-C\equiv C-C_6H_5$	0.00	*1UU1R	CCl_4	25
255	$C_8H_5O_2$		−3.71	*VVR	B	25
256	$C_8H_5O_3$		−3.30	*VOVR	B	25
257	C_8H_5Se		−1.32	*-SE-1UU1R	B	25
258	C_8H_6Br		−1.85	*1U1R DE	B	25
259	C_8H_6Cl	(cis)	−1.68	*YGU1R–C	B	25
260	C_8H_6Cl	(trans)	−1.29	*YGU1R–T	B	25
261	C_8H_6Cl	(cis)	−1.56	*1U1R BG–C	B	25
262	C_8H_6Cl	(trans)	−1.34	*1U1R BG–T	B	25
263	C_8H_6Cl	(trans)	−1.66	*1U1R CG–T	B	25

Appendix I—*Continued*

No.	Formula	R	μ_R (Debye)	WLN[b]	Solvent[c]	Temperature, °C
264	C_8H_6Cl	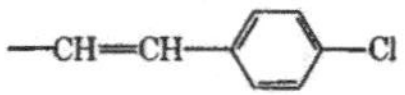	−1.73	*1U1R DG	B	25
265	$C_8H_6Cl_3$	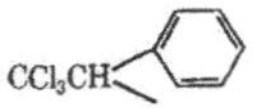	−1.82	*YR&XGGG	B	17
266	C_8H_6F	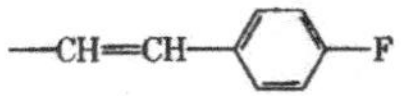	−1.49	*1U1R DF	B	25
267	C_8H_6I	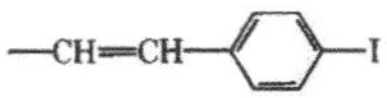	−1.80	*1U1R DI	B	25
268	$C_8H_6NO_2$	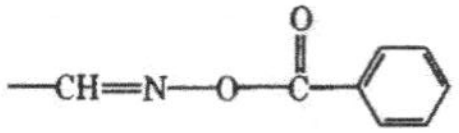	−3.32	*1UNOVR	B	25
269	$C_8H_6NO_2$	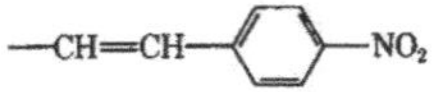	−4.74	*1U1R DNW	B	25
270	C_8H_7	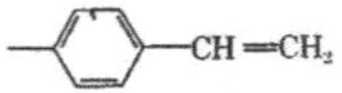	0.64	*R D1U1	B	ns
271	C_8H_7O	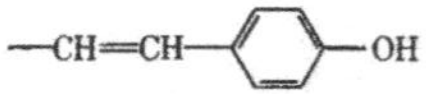	−1.64	*1U1R DQ	B	25
272	C_8H_7O	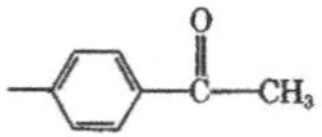	−3.11	*R DV1	B	20–60
273	$C_8H_7O_2$	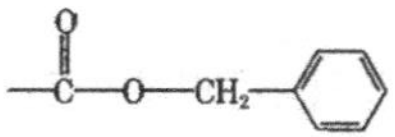	−2.06	*VO1R	B	30
274	$C_8H_7O_3$	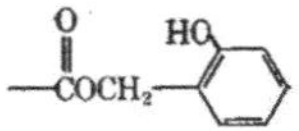	−2.56	*VO1R BQ	B	30
275	C_8H_7SSe	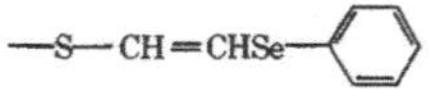	−1.81	*S1U1–SE–R	B	25
276	C_8H_7Se	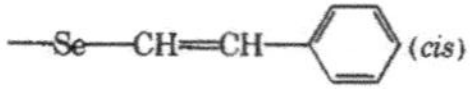	−1.17	*–SE–1U1R–C	B	25

continued

Appendix I-*Continued*

No.	Formula	R	μ_R (Debye)	WLN[b]	Solvent[c]	Temperature, °C
277	C_8H_7Se	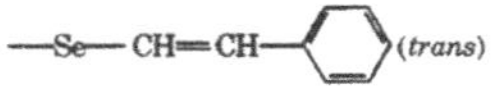 —Se—CH=CH—⟨⟩(*trans*)	−1.06	*-SE-1U1R-T	B	25
278	C_8H_8N	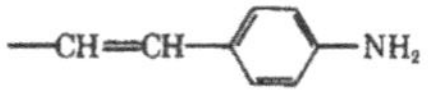 —CH=CH—⟨⟩NH₂	2.06	*1U1R DZ	B	25
279	C_8H_8N	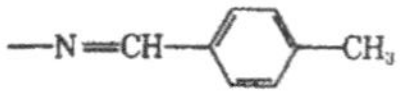 —N=CH—⟨⟩CH₃	−1.93	*NU1R D1	B	25
280	C_8H_8NO	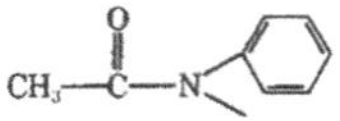 (structure)	−3.61	*NR&V1	B	25
281	C_8H_8NO	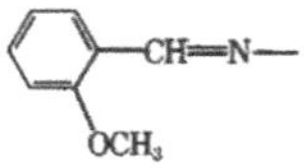 (structure)	−2.87	*NU1R BO1	B	25
282	$C_8H_8N_3O$	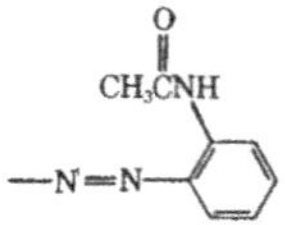 (structure)	−3.47	*NUNR BMV1	B	20
283	$C_8H_8N_3O$	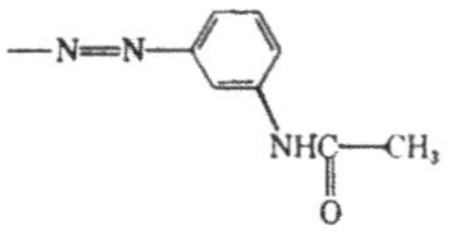 (structure)	−3.71	*NUNR CMV1	B	20
284	$C_8H_8N_3O$	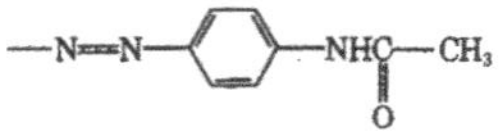 (structure)	−3.72	*NUNR DMV1	B	20
285	C_8H_9OS	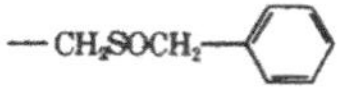 —CH₂SOCH₂—⟨⟩	−3.76	*1SO&1R	B	25
286	$C_8H_9O_2$	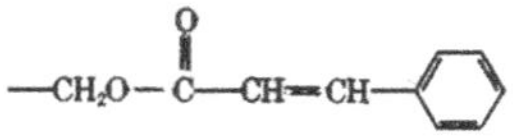 —CH₂O—C(=O)—CH=CH—⟨⟩	−2.27	*1OV1U1R	B	30
287	$C_8H_9O_2S$	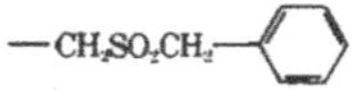 —CH₂SO₂CH₂—⟨⟩	−4.25	*1SW1R	B	25

Appendix I–*Continued*

No.	Formula	R	μ_R (Debye)	WLN[b]	Solvent[c]	Temperature, °C
288	C_8H_9S	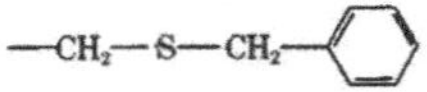 —CH_2—S—CH_2— (phenyl)	−1.34	*1S1R	B	25
289a	$C_9H_9S_2$	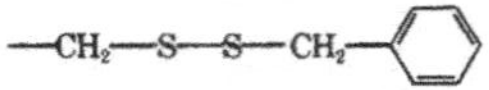 —CH_2—S—S—CH_2— (phenyl)	−1.87	*1SS1R	B	25
289b	$C_8H_{10}NO_2S_2$	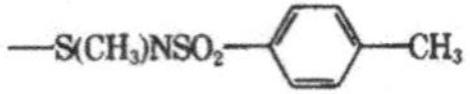 S(CH_3)NSO_2— (phenyl)—CH_3	−7.46	*S1&UNSWR D1	B	20
290	$C_8H_{10}N_3$	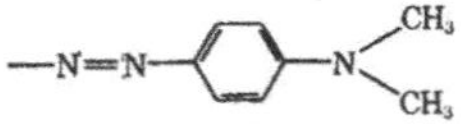 —N=N— (phenyl)—N(CH_3)$_2$	2.82	*NUNR DN1&1	B	20
291	$C_8H_{18}PO$	PO(C_4H_9)$_2$	−4.31	*PO&4&4	B	25
292	C_9H_7OS	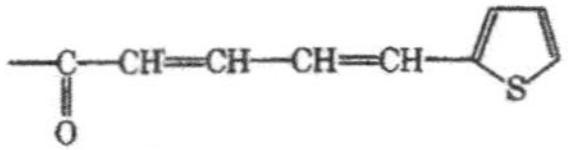 —C(=O)—CH=CH—CH=CH— (thiophene)	−3.25	*V1U2U1–BT5SJ	B	25
293	C_9H_7OS	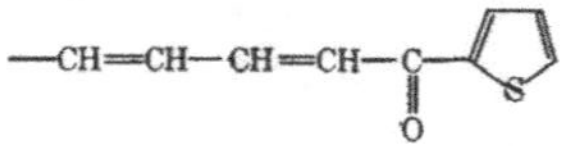 —CH=CH—CH=CH—C(=O)— (thiophene)	−3.50	*1U2U1V–BT5SJ	B	25
294	C_9H_7OS	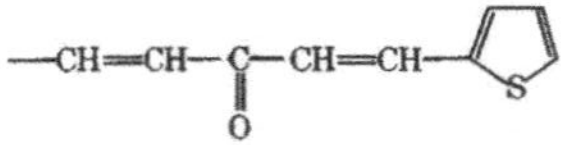 —CH=CH—CH=CH— (thiophene)	−3.21	*1U1V1U1–BT5SJ	B	25
295	$C_9H_7O_2$	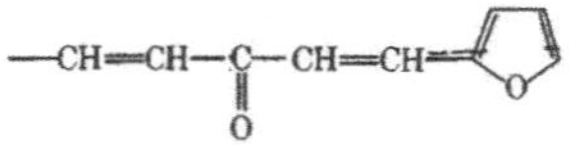 —CH=CH—C(=O)—CH=CH— (furan)	−3.29	*1U1V1U1–BT5OJ	B	25
296	$C_9H_7O_2$	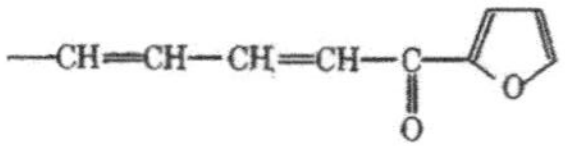 —CH=CH—CH=CH—C(=O)— (furan)	−3.27	*1U2U1V–BT5OJ	B	25
297	$C_9H_7O_4$	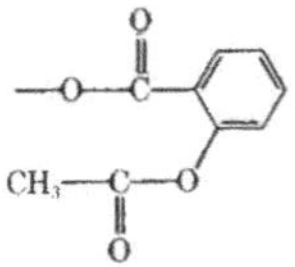 (benzene ring with C(=O)—O—CH_3 and O—C(=O) substituents)	−2.54	*OVR BOV1	B	25

351

Appendix I—*Continued*

No.	Formula	R	μ_R (Debye)	WLN[b]	Solvent[c]	Temperature, °C
298	C_9H_9	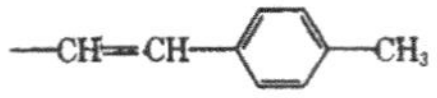 $-CH=CH-$⟨C₆H₄⟩$-CH_3$	0.59	*1U1R D1	B	25
299	C_9H_9O	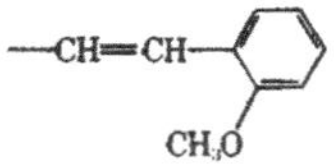 $-CH=CH-$⟨ring with CH_3O⟩	-1.05	*1U1R DO1	B	25
300	C_9H_9O	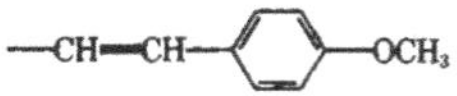 $-CH=CH-$⟨C₆H₄⟩$-OCH_3$	-1.45 -1.13	*1U1R DO1	B	25
301	$C_9H_9O_2$	$-CH_2-O-\overset{O}{\overset{\|}{C}}-CH_2-$⟨phenyl⟩	-1.97	*1OV1R	B	24
302	$C_{10}H_{11}O$	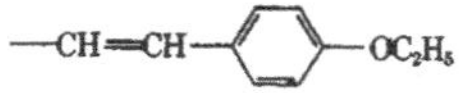 $-CH=CH-$⟨C₆H₄⟩$-OC_2H_5$	-1.66	*1U1R DO2	B	25
303	$C_{10}H_{12}N$	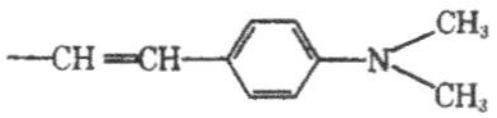 $-CH=CH-$⟨C₆H₄⟩$-N\langle^{CH_3}_{CH_3}$	2.27	*1U1R DN1&1	B	25
304	$C_{12}H_{10}N$	$-N(C_6H_5)_2$	0.70	*NR&R	B	25
305	$C_{12}H_{10}P$	$-P(C_6H_5)_2$	-1.52	*PR&R	B	25
306	$C_{12}H_{10}PO$	$-PO(C_6H_5)_2$	-4.49	*PO&R&R	B	25

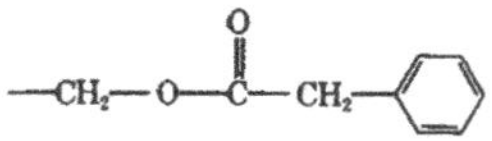

[a]Taken from A. L. McClellan, *Tables of Experimental Dipole Moments*, Vol. 2, Rahara Enterprises, El Cerrito, 1970.

[b]From E. G. Smith, *The Wiswesser Line-Formula Chemical Notation*, McGraw-Hill, New York, 1968.

[c]Solvents: cHx = Cyclohexane, Hx = hexane, B = benzene, D = 1,4-dioxane, L = liquid state, ns = not stated.

[d]V. I. Minkin, O. A. Osipov, and Y. A. Zhdanov, *Dipole Moments in Organic Chemistry*, English Translation by B. J. Hazzard, Plenum, New York, 1970.

[e]O. Exner, V. Jehlieka, and B. Uchytil, Dipole moments and conformation of acetals, *Coll. Czech. Chem Commun.* 33:2826-2871 (1968).

Source: Adapted with permission from *J. Pharm. Sci.* 71(6), 1982; with corrections.

Appendix II . **Aliphatic Dipole Moments and Taft Polar Constants (σ^*)**
$CH_3(CH_2)_n R$, where $n = 0\text{-}6$

No.	Formula	R	n'	μR, Debye	σ^*	WLN	Solvent[a]	Temp., °C
1	BH_2O_2	—B(OH)₂	0.5	-1.16	0.95	*BQQ	G	115
2	Br	—Br	0.0	-1.97	2.84	*E	CCl₄	25
3	Cl	—Cl	0.0	-1.93	2.68	*G	CCl₄	25
4	ClO_2S^b	—SO₂Cl	0.0	-2.28	5.00	*SWG	B	25
5	Cl_2PS	—PSCl₂	0.0	-3.00	3.70	*PS&GG	B	20
6	ClS	—SCl	0.5	-2.00	2.50	*SG	G	—
7	F	—F	0.0	-1.90	3.21	*F	G	NS
8	FO_2S	—SO₂F	0.0	-3.39	4.70	*SWF	B	25
9	F_4P	—PF₄	0.0	-2.55	2.80	*PFFFF	G	31-70
10	I	—I	0.0	-1.79	2.46	*I	CCl₄	25
11	NHCHO	—NHCHO	1.0	-3.86	1.62	*MVH	B	25
12	NCO	—NCO	1.0	-2.81	2.25	*NCO	B	20
13	H_2NO	—NHOH	0.5	-0.80	0.30	*MQ	B	25
14	NH_2	—NH₂	0.0	-1.35	0.62	*Z	B,CHx	25
15	NO	—NO	0.5	-2.30	2.08	*NO	G	NS
16	NO_2	—NO₂	0.5	-3.59	4.25	*NW	G	20
17	NH_2O_2S	—SO₂NH₂	0.5	-4.60	2.61	*SWZ	D	30
18	NO_3	—ONO₂	0.5	-3.08	3.86	*ONW	G	-78
19	N_2H_3	—NHNH₂	0.5	-1.82	0.40	*MZ	G	25
20	N_3	—N=N=N	0.0	-2.17	2.62	*NNN	G	NS
21	OH	—OH	0.0	-1.66	1.55	*Q	B	25
22	SH	—SH	0.0	-1.51	1.68	*SH	B	25
23	SO_2CH_3	—S(O)OCH₃	0.5	-2.83	2.84	*SO&OI	B	25
24	SO_3CH_3	—SO₂CH₃	0.0	-4.16	3.62	*OSWI	D	25
25	$SeCH_3$	—SeCH₃	1.0	-1.41	0.95	*-Se-1	G	NS
26	CCIO	—COCl	0.5	-2.48	1.81	*VG	CCl₄	25
27	CCl_3	—CCl₃	0.0	-1.84	2.65	*XGGG	CHx	25
28	CF_3	—CF₃	0.0	-1.94	2.61	*XFFF	B	25
29	CN	—CN	0.5	-3.63	3.30	*CN	B	30
30	CNS	—SCN	1.0	-3.89	3.43	*SCN	B	25
31	CNSe	—SeCN	1.5	-3.91	3.61	*-Se-CN	B	25
32	$CHBr_2$	—CHBr₂	0.5	-1.90	1.90	*YEE	CHx	25
33	$CHCl_2$	—CHCl₂	0.5	-1.96	1.94	*YGG	B	25
34	CHO	—CHO	0.5	-2.58	2.15	*VH	B	25
35	CHO_2	—COOH	0.5	-1.65	2.08	*VQ	B	30
36	CH_2Br	—CH₂Br	1.0	-1.97	1.00	*1E	CCl₄	25.
37	CH_2Cl	—CH₂Cl	1.0	-1.93	1.05	*1G	CCl₄	25
38	CH_2I	—CH₂I	1.0	-1.79	0.85	*1I	CCl₄	25
39	CH_2NO	—CONH₂	0.5	-3.73	1.68	*VZ	B	25

354

No.	Formula	R	π'	μR, Debye	σ*	WLN	Solvent[a]	Temp., °C
40	CH_2NO_2	—CH_2NO_2	1.5	-3.29	1.73	*1NW	B	25
41	CH_2ClO	—OCH_2Cl	0.0	-1.90	2.56	*O1G	B	0
42	CH_2SH	—CH_2SH	1.0	-1.52	0.62	*1SH	G	-50
43	CH_3	—CH_3	0.0	0.0	0.0	*1	G	—
44	$CH_3N_2S^b$	—$NHCSNH_2$	0.0	-0.16	1.80	*MYZUS	B	25
45	CH_3O	—OCH_3	0.0	-1.27	1.81	*O1	B	25
46	CH_3OS	—$S(O)CH_3$	0.5	-3.88	2.88	*SO&1	B	20
47	CH_3O_2S	—SO_2CH_3	0.5	-4.26	3.68	*SW1	B	25
48	CH_3O_3S	—SO_2OCH_3	0.0	-4.18	3.62	*SWO1	D	25
49	CH_3S	—SCH_3	0.0	-1.45	1.56	*S1	B	25
50	CH_4N	—$NHCH_3$	0.0	+1.01	-0.81	*M1	G	NS
51	CH_4N	—CH_2NH_2	1.0	-1.35	0.50	*1Z	B,CHx	25
52	C_2H	—C≡CH	0.0	-0.78	1.30	*1UU1	NS	NS
53	C_2HS	—SCCH	0.0	-1.69	2.00	*S1UU1	G	-53
54	C_2H_2Cl	—CHCHCl	1.0	-1.64	0.87	*1U1G	G	20
55	$C_2H_2NO_2$	—$CHCHNO_2$	1.5	-3.99	1.75	*1U1NW	CHx	25
56	$C_2H_2Cl_3$	—CH_2CCl_3	0.5	-1.84	0.75	*1XGGG	CCl₄	25
57	C_2H_2OCl	—$COCH_2Cl$	0.5	-2.27	2.50	*V1G	G	-78
58	C_2H_3	—CH=CH_2	0.0	-0.40	0.56	*1U1	B	25
59	$C_2H_3Br_2$	—$CHBrCH_2Br$	1.0	-1.43	1.38	*YE1E	—	—
60	$C_2H_3Cl_3$	—$CHClCHCl_2$	1.0	-2.07	1.08	*YGYGG	—	31–55
61	$C_2H_3Cl_2$	—CCl_2CH_3	0.5	-2.33	1.53	*XGG1	B	25
62	C_2H_3O	—CH_2CHO	1.5	-2.23	0.62	*1VH	B	30
63	C_2H_3O	—$COCH_3$	0.5	-2.77	1.81	*V1	B	25
64	$C_2H_3O_2$	—CH_2COOH	1.5	-1.68	1.08	*1VQ	B	25
65	$C_2H_3O_2$	—$COOCH_3$	0.5	-1.75	2.00	*VO1	B	25
66	$C_2H_3O_2$	—$OCOCH_3$	0.0	-1.81	2.56	*OV1	B	25
67	C_2H_3S	—SCH=CH_2	0.0	-1.38	1.31	*S1U1	B	25
68	C_2H_4Br	—$CHBrCH_3$	1.0	-2.08	1.25	*YE	CCl₄	25
69	C_2H_4Br	—CH_2CH_2Br	2.0	-1.97	0.44	*2E	CCl₄	25
70	C_2H_4Cl	—$CHClCH_3$	1.0	-2.05	1.00	*YG	CCl₄	25
71	C_2H_4Cl	—CH_2CH_2Cl	2.0	-1.93	0.41	*2G	CCl₄	25
72	C_2H_4I	—CH_2CH_2I	2.0	-1.79	0.41	*2I	B	25
73	C_2H_4NO	—$NHCOCH_3$	1.0	-3.81	1.40	*MV1	CCl₄	25
74	C_2H_4NO	—CH_2CONH_2	1.5	-3.75	0.31	*1VZ	B	25
75	$C_2H_4NO_2$	—$CH_2CH_2NO_2$	2.5	-2.69	0.50	*2NW	B	30
76	C_2H_5	—C_2H_5	0.0	0.0	-0.10	*2	G	NS
77	C_2H_5O	—OC_2H_5	0.0	-1.27	1.68	*O2	B	25
78	C_2H_5O	—CH_2OCH_3	1.0	-1.32	0.66	*1O1	CCl₄	25
79	C_2H_5O	—CH(OH)CH_3	1.0	-1.69	0.46	*YQ	CHx	25

continued

Appendix (continued)

No.	Formula	R	n'	μR, Debye	σ*	WLN	Solvent[a]	Temp., °C
80	C_2H_5O	—CH₂CH₂OH	2.0	-1.66	0.21	*2Q	B	25
81	$C_2H_5O_2S$	—S(O)OC₂H₅	0.5	-2.84	2.84	*SO&O2	B	25
82	$C_2H_5O_2S$	—CH₂SO₂CH₃	1.5	-4.40	1.32	*1SW1	B	25
83	C_2H_6N	—N(CH₃)₂	0.0	+1.26	-0.62	*N1&1	B	25
84	$C_2H_6NO_2S$	—SO₂N(CH₃)₂	0.5	-4.71	2.62	*SWN1&1	B	25
85	$C_2H_6NO_2S$	—NCH₂SO₂CH₃	1.0	-4.71	2.10	*N1&SW1	B	25
86	C_2H_6OP	—PO(CH₃)₂	1.0	-4.20	2.81	*PO&1&1	B	25
87	C_3H_3	—CCCH₃	0.0	-0.84	1.20	*1UU2	G	—
88	C_3H_3	—CH₂CCH	1.0	-0.84	0.81	*2UU1	G	35
89	$C_3H_4F_3$	—CH₂CH₂CF	2.0	-1.94	0.32	*2XFFF	—	—
90	C_3H_4N	—CH₂CH₂CN	2.5	-3.51	0.49	*2CN	CCl₄	20
91	C_3H_5	—C(CH₃)CH₂	0.0	-0.34	0.48	*YU1	B	25
92	C_3H_5	—CH₂CHCH₂	1.0	-0.35	0.0	*2U1	G	-78
93	C_3H_5	—CHCHCH₃	0.0	-0.25	0.36	*1U2	G	NS
94	cyclo-C_3H_5	cyclo-C₃H₅	0.0	-0.14	0.15	*AL3TJ	G	—
95	C_3H_5O	—COC₂H₅	0.5	-2.79	1.61	*V2	B	25
96	C_3H_5O	—CH₂CH₂CHO	2.5	-2.23	0.29	*2VH	B	25
97	C_3H_5O	—CH₂COCH₃	1.5	-2.80	0.62	*1V1	B	25
98	$C_3H_5O_2$	—CH₂OCOCH₃	1.0	-1.84	1.06	*1OV1	B	25
99	$C_3H_5O_2$	—COOC₂H₅	0.5	-1.81	2.26	*V02	B	25
100	$C_3H_5O_2$	—CH₂COOCH₃	1.5	-1.84	1.00	*1V01	B	25
101	C_3H_6NO	—CON(CH₃)₂	0.5	-3.81	1.94	*VN1&1	B	25
102	C_3H_6NO	—NHCOC₂H₅	1.0	-3.55	1.56	*MV2	G	110
103	C_3H_6NO	—N(CH₃)COCH₃	1.0	-3.86	2.25	*N1&V1	B	25
104	C_3H_6NO	—CH₂CH₂CONH₂	2.5	-3.78	0.19	*2VZ	B	25
105	C_3H_6NO	—CH₂NHCOCH₃	2.0	-3.55	0.43	*1MV1	G	110
106	$C_3H_6NO_2$	—NHCOOC₂H₅	1.0	-3.80	1.99	*MV02	/	20
107	$C_3H_6NO_2$	—OCON(CH₃)₂	0.0	-3.80	2.87	*OVN1&1	/	20
108	C_3H_7	—CH(CH₃)₂	0.0	+0.08	-0.19	*Y	/	—
109	C_3H_7	—C₃H₇	0.0	+0.08	-0.12	*3	/	—
110	C_3H_7O	—OCH(CH₃)₂	0.0	-1.32	1.62	*OY	B	25
111	C_3H_7O	—CH₂OC₂H₅	1.0	-1.27	0.58	*102	B	25
112	C_3H_7O	—OC₃H₇	0.0	-1.32	1.68	*03	CCl₄	25
113	C_3H_7O	—CH₂CH(OH)CH₃	2.0	-1.77	0.16	*1YQ	D	25
114	C_3H_7O	—CH₂CH₂OCH₃	2.0	-1.27	0.24	*201	B	25
115	C_3H_7O	—C(OH)(CH₃)₂	1.0	-1.72	0.35	*XQ	D	25
116	C_3H_7S	—SC₃H₇	0.0	-1.63	1.38	*S3	B	25
117	C_3H_7S	—SCH(CH₃)₂	0.0	-1.61	1.49	*SY	B	19
118	$C_3H_7O_2S$	—SO₂CH(CH₃)₂	0.5	-4.50	3.68	*SWY	B	20
119	C_4H_7	—CHCHC₂H₅	0.0	-0.34	0.31	*1U3	B	25

No.	Formula	R	n'	μR, Debye	σ*	WLN	Solvent[a]	Temp., °C
120	C₄H₇	—CHC(CH₃)₂	0.0	−0.34	0.19	*1UY	B	25
121	C₄H₇	—CH₂CHCHCH₃	1.0	−0.34	0.0	*2U2	B	25
122	C₄H₉	—C₄H₉	0.0	0.08	0.25	*4	B	25
123	C₄H₉O	—OC₄H₉	0.0	−1.26	1.68	*04	B	25
124	C₄H₁₀N	—NHC₄H₉	0.0	+1.27	−1.08	*M4	B	20
125	C₄C₁₀O₃P	—OP(OC₂H₅)₂	0.0	−2.88	3.02	*PO&O2&O2	B	25
126	C₅H₉O	—O-cyclo-C₅H₉	0.0	−1.60	1.62	*OAL5TJ	/	25
127	C₅H₁₁	—C₅H₁₁	0.0	0.10	−0.23	*5	/	NS
128	C₅H₁₁O	—O(CH₂)₄CH₃	0.0	−1.32	1.52	*05	B	25
129	C₅H₁₁S	—S(CH₂)₄CH₃	0.0	−1.63	1.35	*S5	D	25
130	C₅H₁₁O₂	—CH(OC₂H₅)₂	0.5	−1.27	1.14	*Y02&O2	B	24
131	C₅H₈O₂	—CH₂CHCOOC₂H₅	1.5	−1.95	1.12	*1U1V02	B	24
132	C₆H₂N₃O₆	—C₆H₂-2,4,6-(NO₂)₃	0.0	−1.19	1.62	*R BNW DNW FNW	B	25
133	C₆H₃Cl₂O	—OC₆H₃-2,4-Cl₂	0.0	−2.77	3.17	*OR BG DG	B	20
134	C₆H₄Br	—C₆H₄—4-Br	1.0	−1.91	0.86	*R DE	B	25
135	C₆H₄Cl	—C₆H₄—2-Cl	1.0	−1.34	1.05	*R BG	HP	20
136	C₆H₄Cl	—C₆H₄—4-Cl	1.0	−1.90	0.92	*R DG	B	25
137	C₆H₄Cl	—C₆H₄—3-Cl	1.0	−1.82	0.85	*R CG	B	25
138	C₆H₄F	—C₆H₄—3-F	1.0	−1.78	0.82	*R CF	B	25
139	C₆H₄F	—C₆H₄—4-F	1.0	−1.78	0.81	*R DF	B	25
140	C₆H₄I	—C₆H₄—4-I	1.0	−1.76	0.87	*R DI	B	25
141	C₆H₄NO₂	—C₆H₄—2-NO₂	1.5	−3.60	1.14	*R BNW	/	9-25
142	C₆H₄NO₂	—C₆H₄—3-NO₂	1.5	−3.40	1.21	*R CNW	B	25
143	C₆H₄NO₂	—C₆H₄—4-NO₂	1.5	−4.43	1.26	*R DNW	B	25
144	C₆H₄BrO	—OC₆H₄—2-Br	0.0	−2.50	2.45	*OR BE	CCl₄	20
145	C₆H₄BrO	—OC₆H₄—3-Br	0.0	−2.05	2.48	*OR CE	B	25
146	C₆H₄BrO	—OC₆H₄—4-Br	0.0	−2.37	2.44	*OR DE	B	25
147	C₆H₄ClO	—OC₆H₄—3-Cl	0.0	−2.06	2.57	*OR CG	B	25
148	C₆H₄ClO	—OC₆H₄—4-Cl	0.0	−2.30	2.62	*OR DG	B	25
149	C₆H₄IO	—OC₆H₄—2-I	0.0	−2.25	2.38	*OR BI	CCl₄	20
150	C₆H₄IO	—OC₆H₄—4-I	0.0	−2.14	2.39	*OR DI	CCl₄	20
151	C₆H₄NO₃	—OC₆H₄—2-NO₂	2.0	−4.05	2.78	*OR BNW	B	30, 50
152	C₆H₄NO₃	—OC₆H₄—3-NO₂	2.0	−4.00	2.76	*OR CNW	B	20
153	C₆H₄NO₃	—OC₆H₄—4-NO₂	2.0	−4.00	2.91	*OR DNW	B, D	30, 60
154	C₆H₄BrS	—SC₆H₄—3-Br	0.0	−1.83	1.84	*SR CE	B	30
155	C₆H₄BrS	—SC₆H₄—4-Br	0.0	−1.80	1.83	*SR DE	B	25
156	C₆H₄ClS	—SC₆H₄—3-Cl	0.0	−1.83	2.02	*SR CG	B	30
157	C₆H₄ClS	—SC₆H₄—4-Cl	0.0	−1.81	1.97	*SR DG	B	30
158	C₆H₄FS	—SC₆H₄—4-F	0.0	−1.64	1.77	*SR DF	B	30
159	C₆H₄NO₂S	—SC₆H₄—2-NO₂	2.5	−4.93	2.47	*SR BNW	B	NS
160	C₆H₄NO₂S	—SC₆H₄—4-NO₂	2.5	−4.43	2.33	*SR DNW	B	25

Continued on next page

357

Appendix (continued)

No.	Formula	R	n′	μR, Debye	σ*	WLM	Solvent[a]	Temp., °C
161	C_6H_4NOSe	—SeC$_6$H$_4$—4-NO$_2$	2.5	−4.38	1.83	*SeR DNW	B	NS
162	$C_6H_4NO_4S$	—SO$_2$C$_6$H$_4$—4-NO$_2$	0.0	−2.80	3.63	*SWR DNW	D	25
163	C_6H_4BrOS	—S(O)C$_6$H$_4$—4-Br	0.5	−3.26	3.14	*SO&R DE	B	20
164	C_6H_4ClOS	—S(O)C$_6$H$_4$—4-Cl	0.5	−3.08	3.14	*SO&R DG	B	20
165	C_6H_5	—C$_6$H$_5$	0.0	−0.38	0.75	*R	CHx	25
166	C_6H_5O	—OC$_6$H$_5$	0.0	−1.38	2.43	*OR	B	25
167	$C_6H_5O_2$	—OC$_6$H$_4$—2-OH	0.0	−2.46	2.60	*OR BQ	B	60
168	$C_6H_5O_2S$	—OSOC$_6$H$_5$	0.0	−3.48	3.25	*OSO&R	B	25
169	$C_6H_5O_2S$	—SO$_2$C$_6$H$_5$	0.5	−4.75	3.55	*SWR	B	25
170	$C_6H_5O_3S$	—OSO$_2$C$_6$H$_5$	0.0	−4.99	3.62	*OSWR	D	25
171	C_6H_5S	—SC$_6$H$_5$	0.0	−1.29	1.87	*SR	B	25
172	C_6H_5S	—C$_6$H$_4$—2-SH	1.0	−1.16	0.72	*R BSH	B	25
173	$C_6H_6NO_2S$	—NHSO$_2$C$_6$H$_5$	1.0	−4.62	1.99	*MSWR	B	25–55
174	C_6H_{11}	—cyclohexyl	0.0	0.00	0.18	*AL6TJ	CCl$_4$	25
175	$C_6H_{11}O$	—O-cyclohexyl	0.0	−1.68	1.81	*O AL6TJ	B	25
176	$C_6H_{13}S$	—S(CH$_3$)$_5$CH$_3$	0.0	−1.56	1.33	*S6	B	25
177	C_7H_4NO	—O—C$_6$H$_4$—4-CN	0.0	−4.39	2.73	*OR DCN	B	20
178	C_7H_4ClO	—OCO(C$_6$H$_4$—4-Cl)	0.0	−1.94	2.63	*OVR DG	B	25
179	$C_7H_4NO_4$	—OCO(C$_6$H$_4$—4-NO$_2$)	0.0	−3.48	2.73	*OVR DNW	B	25
180	C_7H_5O	—CO(C$_6$H$_5$)	0.5	−2.90	2.26	*VR	B	20
181	C_7H_5O	—OCOC$_6$H$_5$	0.0	−1.94	2.57	*OVR	B	20
182	C_7H_5O	—COOC$_6$H$_5$	0.0	−1.69	2.57	*VOR	B	25
183	C_7H_6NO	—CONHC$_6$H$_5$	1.0	−3.62	1.68	*VNR	B	25
184	C_7H_7	—CH$_2$C$_6$H$_5$	1.0	−0.39	0.27	*1R	CHx	25
185	C_7H_7	—C$_6$H$_5$—2-CH$_3$	0.0	−0.54	0.62	*R B1	CHx	25
186	C_7H_7	—C$_6$H$_5$—4-CH$_3$	0.0	−0.10	0.59	*R D1	l	25
187	C_7H_7O	—CH$_2$O—C$_6$H$_5$	1.0	−1.38	0.87	*1OR	B	25

358

Appendix (continued)

No.	Formula	R	π'	μR, Debye	σ*	WLM	Solvent[a]	Temp., °C
188	C_7H_7O	—OC_6H_4—2-CH_3	0.0	−1.09	2.29	*OR B1	B	20
189	C_7H_7O	—OC_6H_4—3-CH_3	0.0	−1.25	2.33	*OR C1	B	20
190	C_7H_7O	—OC_6H_4—4-CH_3	0.0	−1.23	2.30	*OR D1	B	20
191	C_7H_7O	—C_6H_4—4-OCH_3	1.0	−1.23	0.42	*R D01	B	20
192	$C_7H_7O_2$	—OC_6H_4—2-OCH_3	0.0	−1.26	2.29	*OR B01	B	20
193	$C_7H_7O_2$	—OC_6H_4—3-OCH_3	0.0	−1.58	2.42	*OR C01	B	20
194	$C_7H_7O_2$	—OC_6H_4—4-OCH_3	0.0	−1.72	2.32	*OR D01	B	25
195	C_7H_7S	—$SCH_2C_6H_5$	0.0	−1.46	1.56	*S1R	D	30
196	C_7H_7S	—SC_6H_4—3-CH_3	0.0	−1.38	1.89	*SR C1	B	30
197	C_7H_7S	—SC_6H_4—4-CH_3	0.0	−1.49	1.80	*SR D1	B	30
198	C_7H_7OS	—SC_6H_4—3-OCH_3	0.0	−1.74	1.89	*SR C01	B	30
199	C_7H_7OS	—SC_6H_4—4-OCH_3	0.0	−1.98	1.66	*SR D01	B	30
200	$C_7H_7O_2S$	—$S(O)C_6H_4$—4-OCH_3	0.5	−4.24	3.00	*SO&R D01	D	20
201	$C_7H_7O_2S$	—$SO_2C_6H_4$—4-CH_3	0.5	−5.08	3.32	*SWR D1	D	25
202	$C_7H_7S_2$	—SC_6H_4—4-SCH_3	0.0	−1.81	1.69	*SR DS1	B	30
203	C_7H_7Se	—SeC_6H_4—4-CH_3	1.0	−1.46	1.23	*Se- R D1	B	NS
204	$C_7H_{13}O$	—OCH_2—cyclo-C_6H_{11}	0.0	−1.41	1.31	*01 A16TJ	B	20
205	C_8H_7	—CH=CHC_6H_5	0.0	−0.77	0.41	*1U1R	B	25
206	C_8H_9	—$CH(CH_3)C_6H_5$	1.0	−0.40	0.37	*YR	CHx	25
207	$C_8H_7O_2$	—OC_6H_4—4-$COCH_3$	0.0	−3.04	2.91	*OR DV1	B	25
208	C_8H_8NO	—$N(COCH_3)C_6H_5$	1.0	−3.63	1.37	*NR&V1	B	25
209	$C_8H_9O_3S$[b]	—$CH_2OSO_2C_6H_4$—4-CH_3	1.0	−5.30	1.44	*10SWR D1	D	25
210	C_9H_{11}	—C_6H_4—4-$CH(CH_3)_2$	0.0	−0.15	0.56	*R DY	l	25
211	$C_{10}H_{13}$	—C_6H_4—4-$C(CH_3)_3$	0.0	0.0	0.52	*R DX	CCl_4	25
212	$C_{12}H_{10}P$	—$P(C_6H_5)_2$	0.5	−1.39	1.06	*PR&R	B	20
213	$C_{12}H_{10}OP$	—$PO(C_6H_5)_2$	1.0	−4.66	1.71	*PO&R&R	B	20
214	$C_{13}H_{11}S_2$	—$CH(SC_6H_5)_2$	0.5	−1.72	1.56	*YSR&SR	B	25

[a] Key: (B) benzene; (D) dioxane; (G) gaseous phase; (CHx) cyclohexane; (HP) heptane; (l) liquid; (NS) not stated. [b] Omitted in equations 8 and 10.

Source: Adapted from *J. Pharm. Sci.* **73**:553 (1986), with permission from the Publisher.

Index

T - #0166 - 101024 - C0 - 229/152/21 [23] - CB - 9780824776862 - Gloss Lamination